About IFPRI

The International Food Policy Research Institute (IFPRI®) was established in 1975 to identify and analyze alternative national and international strategies and policies for meeting food needs of the developing world on a sustainable basis, with particular emphasis on low-income countries and on the poorer groups in those countries. While the research effort is geared to the precise objective of contributing to the reduction of hunger and malnutrition, the factors involved are many and wide-ranging, requiring analysis of underlying processes and extending beyond a narrowly defined food sector. The Institute's research program reflects worldwide collaboration with governments and private and public institutions interested in increasing food production and improving the equity of its distribution. Research results are disseminated to policymakers, opinion formers, administrators, policy analysts, researchers, and others concerned with national and international food and agricultural policy.

About IFPRI Research Monographs

IFPRI Research Monographs are well-focused, policy-relevant monographs based on original and innovative research conducted at IFPRI. All manuscripts submitted for publication as IFPRI Research Monographs undergo extensive external and internal reviews. Prior to submission to the Publications Review Committee, each manuscript is circulated informally among the author's colleagues. Upon submission to the Committee, the manuscript is reviewed by an IFPRI reviewer and presented in a formal seminar. Three additional reviewers—at least two external to IFPRI and one from the Committee—are selected to review the manuscript. Reviewers are chosen for their familiarity with the country setting. The Committee provides the author its reaction to the reviewers' comments. After revising as necessary, the author resubmits the manuscript to the Committee with a written response to the reviewers' and Committee's comments. The Committee then makes its recommendations on publication of the manuscript to the Director General of IFPRI. With the Director General's approval, the manuscript becomes part of the IFPRI Research Monograph series. The publication series, under the original name of IFPRI Research Reports, began in 1977.

AF810068

From the Ground Up

Impacts of a Pro-Poor Community-Driven Development Project in Nigeria

Ephraim Nkonya, Dayo Phillip, Tewodaj Mogues, John Pender, and Edward Kato

RESEARCH MONOGRAPH

IFPRI®

INTERNATIONAL FOOD POLICY RESEARCH INSTITUTE

sustainable solutions for ending hunger and poverty

Supported by the CGIAR

Copyright © 2010 International Food Policy Research Institute. All rights reserved. Sections of this material may be reproduced for personal and not-for-profit use without the express written permission of but with acknowledgment to IFPRI. To reproduce material contained herein for profit or commercial use requires express written permission. To obtain permission, contact the Communications Division at ifpri-copyright@cgiar.org.

International Food Policy Research Institute
2033 K Street, NW
Washington, D.C. 20006-1002, U.S.A.
Telephone +1-202-862-5600
www.ifpri.org

DOI: 10.2499/9780896291799

Library of Congress Cataloging-in-Publication Data

From the ground up : impacts of a pro-poor community-driven development project in Nigeria / Ephraim Nkonya . . . [et al.].
p. cm. — (IFPRI research monograph)
Includes bibliographical references and index.
ISBN 978-0-89629-179-9 (alk. paper)
1. Agricultural development projects—Nigeria. 2. Economic development—Nigeria. 3. Poverty—Nigeria. I. Nkonya, Ephraim. II. International Food Policy Research Institute. III. Title: Impacts of a pro-poor community-driven development project in Nigeria. IV. Series: IFPRI research monograph.
HD2145.5.F76 2010
338.1′8669—dc22 2010038032

Contents

Tables

Figures

Foreword

The concept of community-driven development (CDD) has become popular because it promises to foster sustainable development projects that are responsive to local priorities, empower local communities, and target poor and vulnerable groups. This research monograph assesses the impacts of the largest agricultural CDD project in Nigeria, Fadama II, which was carried out by the Nigerian government in partnership with the World Bank.

The results show that Fadama II dramatically increased the value of group-owned productive assets, in both absolute value and percentage terms, across all agroecological zones, asset terciles, and genders. Participation in the project also increased the income of beneficiaries by about 60 percent—well above the targeted increase of 20 percent over the six-year term of the project. However, the incomes of beneficiaries in the poorest asset tercile and of female household heads did not grow significantly, at least during the six-year span of the project. And even though the program successfully targeted the poor through its group-owned productive assets component, it did not help beneficiaries to invest in the complementary inputs required to make full use of their productive assets.

The results suggest that the poor need help in accessing affordable rural credit services, which can provide the means to pay for productive assets. Although Fadama II did not focus on this issue, the Fadama III project (which began in late 2008) has addressed this problem.

The approach undertaken in Fadama II is a unique and innovative way to reduce poverty. As IFPRI's research indicates, Fadama II is a success story that can serve as a good example for poverty-reduction programs in Africa and other developing countries.

Shenggen Fan
Director General, IFPRI

Preface

Nigeria is aggressively implementing rural development programs aimed at significantly reducing poverty. The objective of the country's development strategy, the 20:2020 Vision, for example, is to make Nigeria one of the 20 largest economies in the world by the year 2020. These efforts have produced promising results in the past ten years, when the gross domestic product (GDP) grew by an annual average of 7 percent—more than double the 2.6–3.0 percent growth rate the country achieved from 1990 to 1999. The agricultural sector contributed about 47 percent of the GDP growth between 1990 and 2007, the largest contribution from a single sector. Despite this impressive growth, however, poverty in Nigeria remains entrenched. An estimated 54 percent of the Nigerian population lives below the poverty line, suggesting that the majority of the poor are being bypassed by these impressive achievements. What must policymakers do to target the poor?

Government initiatives such as the National Economic Empowerment and Development Strategies (NEEDS and NEEDS II) are currently being designed to empower the poor and vulnerable to escape from poverty and to participate in designing new publicly funded development programs. The Fadama II project was one such program intended to target and empower the poor. Consistent with NEEDS and other poverty-reduction programs, Fadama II followed the community-driven development (CDD) model of empowering communities to plan and implement publicly funded projects. Fadama II has produced impressive outcomes that have helped Nigerian policymakers and development partners to implement poverty-reduction programs. Of particular interest to the Nigerian government is Fadama II's success in targeting the poor. This success provides a lesson not only for Nigeria, but also for other countries designing similar programs.

In recognition of Fadama II's impressive achievement, the project received the 2008 World Bank Africa Award for excellence. The government of Nigeria and the World Bank have also taken the bold step of scaling up this impressive achievement to span the entire country. This decision was influenced by Fadama II's success and the government's desire to target the poor.

Since many programs attempt to reduce poverty in Nigeria, IFPRI used an innovative approach to identify those observed outcomes attributable to Fadama II. The study offers insights into Fadama II's impacts and how the

CDD approach has been able to target the poor—an objective that apparently eluded the programs contributing to the impressive GDP growth of the past decade. The study offers important lessons on a CDD program, its strengths and weaknesses, and how the effectiveness of CDD programs can be enhanced.

We thank IFPRI for conducting and publishing this study. This publication will certainly help policymakers and development partners as they design programs for achieving the 2020 Vision, NEEDS, and other poverty-reduction objectives.

Prof. Sheikh Ahmed Abdullah
Honorable Minister
Federal Ministry of Agriculture and Water Development

Acknowledgments

This report is a result of the excellent collaboration and support we received from the participants in our survey and focus group discussions. We thank the 3,750 household survey respondents and the participants in the 200 focus group discussions for their time and valuable information. We are also indebted to the leaders of the many local governments and economic interest groups that supported our fieldwork.

For their hard work in implementing the fieldwork and data entry and cleaning, we thank the state-level consultants A. C. Iheanacho (Adamawa), J. E. Ochi (Bauchi), O. A. Oladunni (Federal Capital Territory), J. O. Olukosi (Gombe), J. E. Njoku (Imo), D. O. Chickwendu (Kaduna), Bello Faruk Umar (Kebbi), Mure Agbonlahor (Lagos), Adeyemi Kayode (Niger), S. Momoh (Ogun), Wale Oni (Oyo), and Park Idisi (Taraba).

We are also grateful to James Akinwumi for his support and assistance in planning and his professional guidance in conducting this study. Additionally, we appreciate the support of the state-level Fadama II subproject coordinators and staff during data collection and cleaning and report writing.

This study would not have been possible without the crucial support and efficient coordination of Adetunji Oredipe and Frank Idefoh. Their administrative and resource support was crucial in carrying out the study in a very short time. We also thank all staff members of the Fadama II national office for their cordial and efficient support.

We appreciate the financial support we received from the Federal Government of Nigeria and the World Bank. We also thank Simeon Ehui, the Fadama II task team leader, for his professional and collegial support during the preparation and implementation of the study. Further, we extend our appreciation to Thomas Muenzel for his valuable comments on our questionnaire and the draft report, and to B. Daramola and B.Omonona for their reviews and their valuable comments on the draft report. We also thank participants in the medium-term review (MTR) meeting in Abuja, Nigeria, and the Brown Bag seminars conducted by the International Food Policy Research Institute and the World Bank. Any errors in or omissions from this report remain the sole responsibility of the authors.

Acronyms and Abbreviations

ADP	Agricultural Development Program
ATT	average effect of the treatment on the treated
CDD	community-driven development
EIG	economic interest group
FCA	Fadama Community Association
FUA	Fadama User Association
FUG	Fadama User Group
IFPRI	International Food Policy Research Institute
LGA	local government authority
MTR	medium-term review
NEEDS	National Economic Empowerment and Development Strategy
NGO	nongovernmental organization
PAA	productive-asset acquisition
PSM	propensity score matching
SFDO	State Fadama Development Office

Summary

The community-driven development (CDD) approach has become increasingly popular because of its potential to develop projects that are sustainable, are responsive to local priorities, empower communities, and more effectively target poor and vulnerable groups. The purpose of this study is to assess the impacts of Fadama II, which is a CDD project and the largest agricultural project in Nigeria. This study used propensity score matching (PSM) to select 1,728 comparable project beneficiaries and nonbeneficiaries. The study also used double difference methods to compare the impact indicators.

Our results show that Fadama II succeeded in targeting the poor and female household heads in its group-owned productive asset acquisition component. Participation in the project also increased the income of beneficiaries by about 60 percent, which is well above the targeted increase of 20 percent in the six-year period of the project. However, incomes of beneficiaries in the poorest asset tercile and among female household heads did not change significantly in the first year. Thus even though the program successfully targeted the poor through its group-owned productive assets component, there was an unmet need to support beneficiaries to invest in complementary inputs required to make full use of their productive assets. The impact of the program also needs to be verified by monitoring its trend over a longer time—especially among the poorest households and those headed by women. Comparison of nonbeneficiaries residing in Fadama II local government authorities (LGAs) and those outside Fadama II LGAs showed a significant spillover of Fadama II to nonbeneficiaries. The incomes of nonbeneficiaries residing in Fadama II LGAs increased by 18 percent because of spillovers from the Fadama II program through public investments, such as community roads, advisory services, and other services.

We also observed that Fadama II increased the demand for postharvest handling technologies but did not have a significant impact on the demand for financial management and market information. Fadama II reduced the demand for soil fertility management technologies. The decline likely reflects the project's focus on providing postproduction advisory services and suggests the need for the project to increase its support for soil fertility management and thus limit the potential for land degradation resulting from increased agricultural productivity. However, the program increased the probability that participating farmers would adopt soil fertility practices. Fadama II may have prompted farmers to adopt soil fertility management practices to maximize returns to their investments. For example, farmers who invested in irrigation infrastructure were also more likely to use fertilizer and organic manure on the irrigated crops.

Overall, Fadama II achieved its goal of increasing the incomes of the beneficiaries in the first year of its operation. The project also succeeded in targeting the poor and vulnerable in its productive-asset component, even though this targeting did not appear to increase significantly household incomes in the short term among the poorest asset tercile. The unique feature that might have contributed to the significant impact of the project in a short time is its broad approach, which addressed the major constraints limiting the success of CDD projects that focus on only one or two constraints. This result has implications for efforts to reduce poverty in low-income countries. Given that the poor face numerous constraints, a CDD project that simultaneously addresses many constraints will likely build synergies that lead to larger impacts than will a project that addresses only one or two constraints. Thus governments and donors need to pool resources and initiate multipronged CDD programs rather than carry out many isolated projects that are not coordinated with one another.

CHAPTER 1

Introduction

The CDD approach has become a key strategy that is used by both governments and organizations that sponsor development assistance programs (Gillespie 2004; Mansuri and Rao 2004; Platteau 2004). The appeal of CDD arose from recent efforts to (1) empower local communities to participate in decisionmaking and implementation of development programs and (2) promote democracy and decentralization (Manor 1999; Dongier et al. 2001; Kohl 2003; Dasgupta and Beard 2007). Social inclusiveness is one of the key features of CDD programs, for the purpose of fostering involvement of the poor and vulnerable in such interventions. To ensure community participation in decisionmaking, CDD programs are demand driven and support groups or communities rather than individuals (Dongier et al. 2001; Binswanger and Aiyar 2003).

Empirical evidence of the effectiveness of CDD in achieving these objectives is mixed (Mansuri and Rao 2004). Among the interesting questions that have captured the attention of scholars are the sustainability of donor-supported and/or government-managed CDD and its effectiveness in targeting the poor and vulnerable. Khwaja (2001) observed that projects managed by communities were more sustainable than those managed by local governments because of better maintenance. Labonne and Chase (2008) also showed that CDD projects improved trust among group members, increased participation in village assemblies, and generally increased social capital among community members. However, Mosse (1997), Cleaver (1999), and Kleimeer (2000) found that CDD projects that lacked external institutional, financial, and technical support were not sustainable. Similarly, Labonne and Chase (2008) observed that CDD projects led to less investment in other projects and did not have significant impacts on membership in development groups compared to control villages.

Targeting the poor has been found to be one of the challenges of the CDD approach (Farrington and Slater 2006). One argument in favor of CDD asserts that it can improve targeting because CDD projects make better use of local knowledge to define and identify the targeted groups (Mansuri and Rao 2004). A recent study in Senegal showed that a CDD project increased the access of poor families to clean water and health services and increased their consumption expenditures (Arcand and Bassole 2007). This study also noted that village chiefs and local governments played a major role in the placement of CDD projects. Several studies have also shown that CDD programs have been effective in targeting the poor in communities with strong local institutions and fairly homogeneous socioeconomic characteristics (Bardhan and Mookherjee 1999; Conning and Kevane 2002; Platteau 2004; Galasso and Ravallion 2005).

However, elite capture—in which a few individuals in a local community have disproportionate political or economic power and dominate community-based planning, governance, and benefits from community-based programs—remains one of the major challenges of the CDD approach (Dasgupta and Beard 2007). Studies have identified cases of elite capture and failure to empower local communities to participate in development programs. Platteau (2004) observed that a large share of financial support by a foreign nongovernmental organization (NGO) to farmer organizations in western Africa was appropriated by local leaders. Consistent with Ostrom (1990), Platteau (2004) also observed that elite capture is a common problem for many donor-funded projects that support local communities with weak local institutions. Elite capture has even been reported in communities with strong democratic institutions. Dasgupta and Beard (2007) observed that communities with democratic institutions in Indonesia restricted allocation of resources to their poorest members. Surprisingly, this study also observed better targeting of the poor in communities controlled by elites. Based on this observation, Dasgupta and Beard (2007) conclude that there is a difference between elite capture and elite control; in the second case, only decisions are controlled by elites, but resource allocation is targeted to the poor.

Heterogeneity also leads to elite capture and ineffective focus on the poor. Conning and Kevane (2002) observed that the ability of CDD projects to target the poor in heterogeneous communities with high social inequality was worse than that of externally managed programs, whereas the opposite was true in egalitarian communities with open and transparent systems of decisionmaking.

Our study assessed the impact of the CDD project Fadama II, which is the largest agricultural project in Nigeria. Fadama II aims to reduce poverty by supporting communities to acquire infrastructure and productive assets, providing demand-driven advisory services, increasing the capacity of communities to manage economic activities, and reducing conflicts among resource users. This report evaluates the impact of the project on income poverty,[1] access to productive assets, and provision of demand-driven advisory services. It does not evaluate how the project affected the capacity to resolve conflicts among users of *fadama* resources and the capacity of beneficiaries to manage CDD projects.[2] In this report, we also examine whether the project succeeded in targeting the poor and the vulnerable through its poverty-reduction efforts and productive-asset acquisitions.

Chapter 2 supplies background on Fadama II and describes how it applied the CDD approach in its design and implementation. Chapter 3 briefly reviews the initial accomplishments of the project identified by the MTR and discusses what our study contributes beyond this review. Chapter 4 discusses the methods of data collection and analysis used in the study. Chapter 5 reports the initial impacts of Fadama II on productive-asset acquisition and household income; Chapter 6 assesses the project's effects on demand for and use of advisory services. The final chapter draws conclusions and discusses the policy implications of the study findings, with an emphasis on strategies that can be used to ensure sustainability of similar projects and improve targeting to the poor and vulnerable.

---

[1] Income poverty is the most common measure of poverty. It uses income as the indicator of poverty. For example, individuals with incomes of less than US$1 a day are regarded as poor by many studies and reports (see World Bank 2007c).

[2] Fadama is a Hausa word for low-lying flood plains, usually with easily accessible shallow groundwater. Fadama are typically waterlogged during the rainy season but retain moisture during the dry season. These areas are considered to have high potential for economic development through appropriate investments in infrastructure, household assets, and technical assistance.

CHAPTER 2

Fadama II: Background and Approach

Background

Fadama II is a follow-up to Fadama I (phase I of the National Fadama Development Project), which was implemented during 1993-99.[1] Fadama I focused mainly on crop production and largely neglected support of postproduction activities, such as commodity processing, storage, and marketing. The emphasis of Fadama I was on providing boreholes and pumps to crop farmers through simple credit arrangements. It aimed to boost aggregate crop output. Fadama I worked with Fadama User Associations (FUAs), which the states used mainly to recover loans and decide on locations to develop water infrastructure.

The design of Fadama I did not support rural infrastructure development and failed to consider other resource users, such as livestock producers, fisherfolk, pastoralists, and hunters. The focus on crop producers contributed to increased conflicts among the diverse types of users of *fadama* resources. In addition, enhanced crop production increased the surplus, but the project did not support postharvest technology, contributing to reduced crop prices and greater storage losses.

Fadama II was first implemented in 2005 and operated in 12 states, 9 of which were Fadama I states (Bauchi, the Federal Capital Territory, Kaduna, Kebbi, Lagos, Niger, Ogun, Oyo, and Taraba) and three that were not (Adamawa, Gombe, and Imo).[2] Fadama II seeks to address the shortcomings of Fadama I by shifting from a top-down and supply-driven public sector development pro-

[1] Fadama I operated in 25 states in Nigeria, of which 9 are also covered by Fadama II. The Fadama I states were Bauchi, Jigawa, Kano, Kebbi, and Sokoto in the north; Benue, Federal Capital Territory, Kogi, Niger, Plateau, and Taraba in the middle belt; and Abia, Akwa Ibom, Anambra, Cross River, Delta, Edo, Enugu, Imo, Lagos, Ogun, Ondo, Oyo, Osun, and Rivers in the south.

[2] We discuss the implications of Fadama I and Fadama II sharing some states in Chapter 4. The projects in Bauchi, Benue, the Federal Capital Territory, Kebbi, Lagos, Niger, Ogun, Oyo, and Taraba receive World Bank support. In another six states—Borno, Jigawa, Katsina, Kogi, Kwara, and Plateau—a version of the project was also implemented with financial support from the African Development Bank.

gram to the CDD approach. Fadama II also includes other *fadama* resource users that the first project had ignored. As discussed later in this chapter, Fadama II also supports activities and services other than production.

Consistent with the CDD approach, project activities are centered on Fadama User Groups (FUGs) and Fadama Community Associations (FCAs). An FUG comprises *fadama* users with a common economic interest and is therefore a type of economic interest group (EIG). The FUGs also included groups that are not related to Fadama resources. For example, beneficiaries formed groups around common nonfarm activities, such as the manufacture of women's apparel and shoe cobbling.

FCAs are the associations of FUGs operating in a given area. Each FCA designs and oversees the implementation of a local development plan, which becomes the blueprint of Fadama II and the development project in that FCA. The major productive sectors that Fadama II supports include crops, livestock, agroforestry, fishing, and fish farming. Addressing one of the weaknesses of Fadama I, Fadama II also supports postproduction activities that are closely linked to the project's productive activities. These include agroprocessing enterprises and rural marketing service providers. As part of its targeting strategy, Fadama II gives special preference to groups of youth, female household heads (especially widows), physically challenged persons, the elderly, and people with HIV/AIDS. Targeted groups can belong to any of the productive or service sectors supported by the project. Because Fadama II uses the CDD approach, beneficiaries are given the chance to choose the kind of activities they want to pursue under the project. However, there are some activities that the project does not support, such as those that could lead to degradation of natural resources or large-scale changes in land use (NFDO 2006). The project also does not support social services, such as building schools and clinics. Under the CDD approach of Fadama II, all users of *fadama* resources are encouraged to develop participatory and socially inclusive local development plans.

Approach

Selection Criteria for Participating States and Beneficiaries

Fadama II was designed to operate for six years (2004–10) with the goal of contributing to poverty reduction in Nigeria. Actual implementation did not begin until September 2005, however. Even though the project focuses on states with significant *fadama* areas, it also operates in other states. For example, Lagos is an urban state but was selected as one of the 12 Fadama II states. Selection of the states was based on their readiness to manage Fadama II. States were also required to prepare Fadama II development programs that target the poor and vulnerable and ensure that the proposed projects do not lead to environmental degradation. States were also supposed to open special bank accounts and deposit an initial amount of money to show their commitment and readiness to manage the program. Assessment and selection of states was done at the federal level.

The local government authorities (LGAs) in each state were selected by the Agricultural Development Project Executive Committee using the following criteria:

1. The regional authorities must be interested in the project and committed to paying counterpart funds on monthly basis.
2. Active EIGs must be committed to establishing FCAs and making a detailed assessment of existing Fadama infrastructure.
3. Two-thirds of the total membership of the Local Fadama Development Committee must be representatives of FCAs and civil society (the Committees plan and manage Fadama II activities at the local level).
4. At least 20 percent of the membership of each Local Fadama Development Committee must be women.
5. At least two qualified staff must be hired to manage Fadama development projects.

Under each LGA, an FCA was formed. These were the umbrella organizations formed by individual EIGs, to which the EIGs submitted applications for support. The FCA selection criteria were as follows:

1. The FCA must be legally recognized by the LGA. Recognition of the FCA is based on having a constitution and an executive committee. Members of the FCA should also be from the same LGA.
2. The FCA must democratically elect leaders of Fadama II subprojects, consisting of at least a chairperson, secretary, and a treasurer.
3. The FCA must have a bank or savings account that is in good standing.
4. The FCA must be committed to a socially inclusive process of Fadama development.
5. The FCA must be commited to paying counterpart contributions for the project.
6. The FCA must supply a written commitment to comply with the project guidelines.

The prospective Fadama II FUGs and EIGs were required to meet the following criteria:

1. The group must have members who come from 20-40 households and who join the FUG voluntarily. The FUG should also be endorsed by the LGA as eligible for Fadama II support. Members have to be from the same LGA.

2. The group must be recognized as a legal civil association with a group constitution and democratically elected leaders consisting of at least a chairperson, secretary, and treasurer.
3. The group must have an active bank or savings account that is in good standing.
4. The group must supply a written commitment to a socially inclusive approach and to compliance with project guidelines and agreements.
5. The group must express interest in and commitment to the project and must apply for Fadama II support.
6. The group must make regular payment of the counterpart funds on a monthly basis.
7. The group must supply evidence of operational and active participation in an FCA or other organized EIG.

These criteria demonstrate that less-organized communities and those not in groups could not benefit from Fadama II subprojects. It is also likely that people living in remote areas where banking services are limited are less likely to benefit from Fadama II.[3] The requirement to pay counterpart funds is also a barrier to the poor, particularly in the case of expensive projects, such as rural infrastructure, large processing machines, and irrigation infrastructure. To ensure the inclusion of women, the FCAs were required to have at least 20 percent female beneficiaries.

Project Goals and Components

The project set a target of 50 percent of all participating *fadama* resource users increasing their average real incomes by at least 20 percent compared with the baseline. The following five components were designed to achieve this goal:

1. *Rural infrastructure investment* to support creation of economic infrastructure and local public goods that would improve the productivity of households using *fadama* resources. Under this component, beneficiaries were required to pay 10 percent of the costs of constructing rural infrastructure, such as rural roads, culverts, market stalls, cold storage, boreholes, and irrigation infrastructure.[4]
2. *Pilot productive-asset acquisition support* to enhance the productivity and income of *fadama* resource users by facilitating the acquisition of produc-

---

[3] However, as discussed in Chapter 5, distance to all-weather roads did not have a significant impact on the probability of participating in Fadama II.

[4] Later we discuss some of the challenges encountered in collecting these co-payments and the subsequent adjustments made to these requirements.

tive assets by individuals or FUGs. Under this component, *fadama* resource users were required to pay 30 percent of the cost of the productive assets acquired.

3. *Demand-responsive advisory services* to support advisory services that will enable *fadama* resource users to adopt output-enhancing techniques and more profitable marketing practices in their enterprises. Beneficiaries were required to pay 10 percent of the cost of providing these services.
4. *Capacity building* to increase the ability of beneficiaries to assess their needs, participate in planning, and implement and manage economic activities, and to increase the capacity of the project coordinators to conduct monitoring and evaluation. Capacity-building support was provided through trained facilitators. In addition, FUG members were trained to negotiate and manage contracts and conduct basic financial analysis.
5. *Conflict resolution* to address one of the shortcomings of Fadama I by increasing the capacity of FUGs to manage conflicts, which were particularly serious and frequent between pastoralists and crop farmers. More than 98 percent of conflicts among *fadama* resource users were between pastoralists and farmers (Schoen, Hassan, and Okoli 2002). The project set an objective of reducing the number of conflicts by 50 percent by 2010.

Because we evaluated the progress of the project and its income impacts after only one full year of implementation, this study should not be considered a final assessment of Fadama II. Rather, it is a quantitative evaluation of initial progress and a potentially useful baseline against which to measure future results.

CHAPTER 3

Past Studies on the Effects of Fadama II

This chapter describes the progress of Fadama II implementation based mainly on the MTR completed in May 2007 (World Bank 2007a). Although this MTR assessed many aspects of the implementation of Fadama II, here we focus on the outcomes analyzed in our study.

Advisory Services

Fadama II has implemented a pluralistic advisory service in which both private and public entities provide services and funds. Advisory service providers are largely private, with only 5 percent of the services offered by public providers. However, funding of the advisory services is mainly public: project beneficiaries pay 10 percent of the cost and the project pays 90 percent. Thus Fadama II has created a foundation for developing demand-driven advisory services using a pluralistic approach, which is an important step in establishing sustainable services.

The MTR states that the advisory service component achieved most of its objectives, although it is not clear how those achievements were measured. For example, the report states that 1,700 advisory services were provided to 1,026 FUGs. However, that achievement affected only 12 percent of the 8,577 FUGs. It is not clear why about 88 percent of the FUGs did not receive advisory services. The MTR also observed collusion between advisory service providers and FCA/FUG officials. This collusion has compromised the independent recruitment of providers and serves as one example of elite capture in CCD projects in developing countries (Mansuri and Rao 2004).

Support for Pilot Productive-Asset Acquisition

According to the MTR, the pilot productive-asset acquisition (PAA) component was readily accepted by beneficiaries, because they obtained tangible near-term benefits from the project, whereas with other components, like infrastructure development, capacity building, or advisory services, the impacts are not as readily felt (World Bank 2007a). A total of 7,511 subprojects were undertaken in the PAA component, representing 67 percent of the subprojects undertaken in all components of Fadama II. At the time of the MTR, at least 67 percent of all PAA subprojects under the local development plans had been completed, and 27 percent were ongoing. Thus 94 percent of PAA subprojects have been funded and almost completed, indicating the effectiveness of and high demand for this component by *fadama* resource users. In the second section of Chapter 5 we quantify the value of productive assets acquired and measure the impact of Fadama II on PAA across several comparison groups.

Rural Infrastructure

The MTR reports that 2,817 rural infrastructure projects had been initiated, 63 percent of which were completed (World Bank 2007a). This significant achievement is likely a result of the large amount of matching funds paid by the project (90 percent), a level that certainly contributed to the fast acceptance and implementation of the rural-infrastructure component. An important issue is the sustainability of the component after the project ends. The commitment of beneficiary communities to adequately maintain the infrastructure using their own resources is not yet clear but might have been undermined by the large contribution from the project. In addition, the 90 percent contribution provided by the project will be costly to replicate in other areas.

Contribution of This Study

In general, the MTR indicates that the accomplishments of Fadama II have been positive for all components except monitoring and evaluation, which was rated as marginally satisfactory (World Bank 2007a). However, the MTR did not quantify the impacts of the project on community and household wealth or income or other expected outcomes of the project. Also, the approach used in the MTR did not control for factors outside the project that could affect outcomes. The major contribution of this study is its approach of investigating counterfactual nonproject communities and households as well as project beneficiaries, allowing better attribution of outcomes to the project. This approach is important for evaluating not only Fadama II but also many other projects whose results are studied without using comparison groups (Mansuri and Rao 2004).

This study uses quasi-experimental and econometric methods to control for factors that could affect project outcomes. The main focus of the study is on quantifying the impacts of the project on poverty reduction, which is the major objective of Fadama II. We do this by examining the impact of Fadama II on the acquisition of productive assets, income, rural infrastructure, and advisory services. The analysis in this report is based on the household survey data only.[1]

[1] Other reports analyze particular impacts or components of Fadama II, including effects on conflict reduction, capacity building, communication and advisory services, and rural infrastructure (Arokoyo 2007; Gbenga 2007; Yahaya 2007). The reports on advisory services and rural infrastructure do not use the survey data used in this study but instead use secondary data and primary data collected by different methods.

CHAPTER 4

Methodological Framework

Study Area

This study was conducted in the 12 states participating in Fadama II. As shown in Figure 4.1, the 12 states lie in three major agroecological zones (Maziya-Dixon et al. 2004): humid forest (Imo, Lagos, and Ogun), moist savannah (Adamawa, the Federal Capital Territory, Oyo, and Taraba), and dry savannah (Bauchi, Gombe, Kaduna, Kebbi, and Niger). In each participating state, the project was implemented in 10 selected LGAs.

Sample Selection

Household Survey

To analyze the impact of Fadama II on beneficiaries and the spillover of benefits to nonparticipants living in Fadama II communities, we divided the sampling frame into three strata: (1) households with at least one member participating directly in the project, (2) households living in Fadama II communities but not directly participating in the project (although they might benefit indirectly), and (3) households living in *fadama* resource areas outside the Fadama II LGAs but with socioeconomic and biophysical characteristics comparable to Fadama II communities in the same state. We expected nonbeneficiaries living in communities participating in Fadama II to be affected by spillovers, such as construction of rural infrastructure and provision of advisory services. For example, project participants living in a Fadama II community that built a culvert could use the same road to transport their produce, and information about new technologies provided by the Fadama II advisory services might be shared with nonparticipants.

This stratification was designed to allow for estimation of the direct and indirect effects of Fadama II. By comparing project outcomes for direct beneficiaries with outcomes for similar (in terms of initial productive-asset endowments, education, and so forth) nonparticipating households in the same communities, we obtained an estimate of the direct impacts of Fadama II participation. Because nonparticipating households in the Fadama II com-

Figure 4.1 Map of Nigeria showing states that participated in the Fadama programs

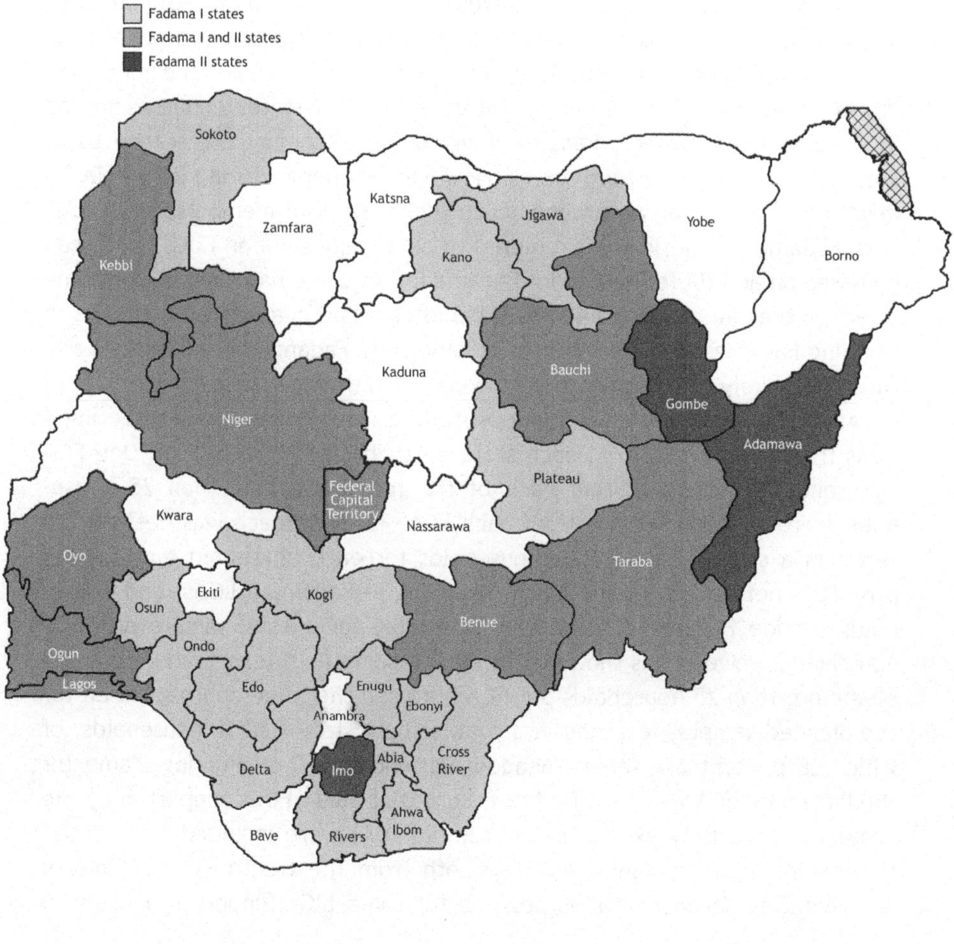

munities may have benefited from spillover effects, this comparison does not provide an estimate of the full impact of the project. Comparing Fadama II beneficiaries to similar households in similar communities not included in the project provides a better estimate of the total impact of the project on beneficiaries (assuming that spillovers do not affect households in the communities outside the project in the relatively short time frame of the study). Likewise, comparing nonparticipants in Fadama II communities with similar households in communities outside the project provides an estimate of the impact of spillover effects on nonparticipants in project communities.

As with Fadama I, the selection of states to participate in Fadama II was not random. The 12 participating states and the local *fadama* resource areas where the project operated were selected by the Government of Nigeria in collaboration with the World Bank.[1] Purposive sampling is common with many government-funded programs in developing countries (Duflo, Glennerster, and Kremer 2006). This process introduces a selection bias and weakens the external validity of our results. Most of the states chosen were in the humid and dry savannah zones. As previously stated, 9 of the 12 states also participated in the Fadama I project. Fadama II did not give special preference to or bias against Fadama I beneficiaries. However, former Fadama I beneficiaries might have derived an advantage because of their prior membership in FUAs. Each Fadama II beneficiary is required to be a member of an FUG, which can be based on an FUA formed under Fadama I. This could have introduced some selection bias in sampling Fadama II beneficiaries, because FUA members in the nine Fadama I states were more likely to be Fadama II beneficiaries and thus more likely to be sampled than were non-FUA members.[2]

At the LGA level, the sampling procedure involved randomly selecting 4 LGAs from among the 10 in each state participating in Fadama II. One FCA was randomly selected from each of the four LGAs, and then 25 households were randomly drawn from each FCA. This approach was designed to result in a sample size of 100 households for each household type (direct project beneficiaries, nonbeneficiaries living in Fadama II LGAs, and households outside Fadama II LGAs) in each state, for a total sample of 3,600 households. However, as shown in Table 4.1, some field teams randomly sampled more than 25 households per FCA but used the same approach used for the planned sample, resulting in a total sample size of 3,750 households, of which 28 percent are female-headed households.[3] The sampling frame for the Fadama II FCA was stratified to ensure that all 14 FUGs supported by the project (where they existed in the sample FCA) were included in the list.[4] Households were randomly selected both from the treatment and control communities. Given that it is possible for some EIGs funded by Fadama II

[1] The project planners did not take randomization into account when designing Fadama II. This study was initiated about a year after the project started, and so did not influence the design of the project.

[2] The double-difference estimator that we used in the impact evaluation helps to address the effects of such differences in initial conditions by subtracting out their additive fixed effects. However, it does not completely solve the potential problem of selection bias, because the impacts of Fadama II may interact with participation in Fadama I. These issues are discussed further in Chapter 5.

[3] At the national level, 18 percent of rural household heads are female (NBS 2005). However, Fadama II targeted women—hence the overrepresentation in the sample households.

[4] The 14 FUGs were crop farmers, fisherfolk, pastoralists, livestock farmers, hunters, gatherers, agroforesters, agroprocessors, service providers, elderly persons, widows, people living with HIV/AIDS, unemployed youths, and physically challenged persons.

Table 4.1 Planned and realized household sampling

Household type	Sample size (number of households) Planned	Actual
FII beneficiary	1,200	1,281
Nonbeneficiary in FII LGA	1,200	1,240
Nonbeneficiary outside FII LGA	1,200	1,229
Total	3,600	3,750

Note: FII, Fadama II; LGA, local government authority.

to have more than one household member belonging to it, sampling at the household level ensured that each household in a community had an equal chance of being selected. Each sampled respondent—which was usually the household head—supplied data for the entire household. Individual-specific information in the questionnaire was asked with respect to the household head. The sampling frame of the household survey was also stratified by the gender of the household head, ensuring that a quarter of the households from each FCA were female-headed households.

Selection of nonbeneficiaries living in and outside Fadama II LGAs followed the same procedure as just described. However, the FUG listed depended on the availability of EIGs comparable to those in the Fadama II. Similarly, 25 percent of the sample consisted of female-headed households.

Focus Group Discussion

The main aim of the community-level focus group discussion was to discuss community organizations, rural infrastructure, and conflicts over resource use. The sampling procedure of communities closely followed the household-level approach. However, only two strata were used: Fadama II and non-Fadama II LGAs. It was not feasible to establish spillover effects by selecting communities in the neighborhood of Fadama II communities (as in the household survey sample), because some FCAs covered more than one village. Respondent groups among the Fadama II beneficiaries were chosen from a randomly selected group of 10–25 individuals who did not participate in the household survey. The individuals were selected from the four FCAs sampled in the household survey. The selected individuals were then separated into two focus groups for the first three LGAs and into four focus groups in the fourth LGA. This process resulted in 10 discussion groups. The same procedure was used to select groups from the non-Fadama II communities; that is, the same LGAs selected for the household survey were used to select 10 groups of *fadama* resource users who do not benefit from Fadama II. The EIGs selected were closely related to those supported by Fadama II. However, establishing those groups was difficult, because the EIGs in non-Fadama II communities are generally not well organized.

Survey Instruments and Data Collection

A structured survey instrument was used for the household survey. The focus group discussion was semistructured; it included both structured questions and discussion guidelines. Structured questions were used to determine the extent of conflict resolution among *fadama* resource users and changes in rural infrastructure. Guidelines were used to direct qualitative discussions about what factors led to conflict resolution and infrastructure changes, how they have affected livelihoods in the community, and what needs to be done in the future. Each of these instruments was developed through meetings, discussions, and pretesting. In each state, the state team leader was responsible for the administration of each type of survey instrument. However, the interviews were carried out by trained enumerators under the supervision of group team leaders. In each state, group team leaders reported to the state team leader at the end of each survey day.

The double-difference analysis used in this study (explained further in the next section) requires baseline data of good quality. Because the baseline survey for Fadama II had some deficiencies (Faye and Sutherland 2006), we collected baseline data for Fadama II using recall information. The project was implemented in September 2005, only slightly more than a year before the survey was conducted; therefore, we expected respondents to be able to remember the baseline data required for two years before the survey—that is, for the crop years October 2004-September 2005 (2004-05) and October 2005-September 2006 (2005-06). This recall information included data on household composition and size, major productive assets, and major components of household income. Household respondents had no difficulty recalling changes in household composition, size, or major productive assets since October 2004, but the recall of income components posed some difficulties. However, because income was not used as an explanatory variable in the analysis (unlike prior household composition and assets) but only as a dependent variable, the potential for measurement error in that variable was of less concern, although it increased uncertainty and reduced the statistical power of the estimates.[5]

[5] In econometric analysis, measurement error in a dependent variable increases the uncertainty of the estimates but causes no bias (as long as the error is not correlated with the explanatory variables), whereas measurement error in an explanatory variable does cause a bias (Greene

The two crop years, 2004-05 and 2005-06, were comparable in terms of rainfall. Both years were reported to be normal in terms of agricultural production, even though production of cereals and tubers for 2004-05 was 8 percent lower than for 2005-06—mainly because of delayed delivery of inputs (FEWSNET 2006). Fortunately, however, the unfavorable production conditions affected both treatment and control groups—thereby netting out the effect on the 2004-05 season.

Data Analysis

Impact assessment studies face three interrelated challenges: establishing a viable counterfactual (the predicted outcome in the absence of the intervention —that is, what would have happened to the beneficiaries had they not participated in the project); attributing the impact to an intervention; and coping with long and unpredictable lag times (Alston and Pardey 2001; Salter and Martin 2001). If a project's outcome indicator is household income, the average impact of the project on its beneficiaries (referred to in the impact assessment literature as the average effect of the treatment on the treated [ATT]) is defined as the difference between the expected income earned by project beneficiaries while participating in the project and the expected income they would have received if they had not participated in the project:

$$ATT = E(Y_1 | p = 1) - E(Y_0 | p = 1), \quad (1)$$

where ATT is ATT, p indicates participation in the project ($p = 1$ if the subject participated in the project, and $p = 0$ if the subject did not participate); Y_1 is the outcome (household income, in this example) of the project beneficiary after participation in the project; and Y_0 is the outcome of the same beneficiary if he or she had not participated in the project.

Unfortunately, we cannot observe the counterfactual income of the beneficiaries had they not participated in the project, that is, $E(Y_0 | p = 1)$. Simply comparing incomes of households participating in the project with those not participating could result in serious biases, because the two groups may be quite different and thus likely to have different incomes regardless of their participation. For example, adding and subtracting $E(Y_0 | p = 0)$ on the right side of equation (1) results in:

$$ATT = [E(Y_1 | p = 1) - E(Y_0 | p = 0)] - [E(Y_0 | p = 1) - E(Y_0 | p = 0)]. \quad (2)$$

2003). We believe that similar principles apply to the results of PSM (the quasi-experimental approach used in this study), although we have not seen specific articles on this issue in the relatively recent literature on this approach.

The expression in the first set of square brackets is observable, because it is the difference between the incomes of the beneficiaries and nonbeneficiaries. The second bracketed expression is unobservable, because $E(Y_0 | p = 1)$ is unobservable and thus represents the bias resulting from estimating ATT as the first expression. This bias results because the incomes that nonbeneficiaries receive without the project may not be equal to the incomes that beneficiaries would have received without the project; that is, $E(Y_0 | p = 1)$ may not equal $E(Y_0 | p = 0)$.

Two common sources of bias are (1) project placement or targeting bias, in which the location or target population of the project is not random (such as when some subprojects of Fadama II are targeted to the poor and vulnerable, so that wealthier groups do not have an equal chance of participating), and (2) self-selection bias, in which households choose whether to participate and thus may be different in their experiences, endowments, and abilities.[6]

The most widely accepted method to address these biases is to use an experimental approach to construct an estimate of the counterfactual situation by randomly assigning households to treatment (beneficiary) and control (nonbeneficiary) groups. Random assignment ensures that both groups are statistically similar (that is, that they are drawn from the same distribution) in both observable and unobservable characteristics, thus avoiding project placement and self-selection biases. Such an approach is not feasible in the present study, because project placement and participation decisions were already made before the design of the study and were probably not random. The notion of random assignment also conflicts with the nature of this CDD project, in which communities and households make their own decisions about whether to participate and what activities they will pursue, thus limiting the ability to use a randomized approach at the outset.

Various quasi-experimental and nonexperimental methods have been used to address the bias problem (for details, see Rosenbaum and Rubin 1983; Heckman, Ichimura, and Todd 1998; Heckman et al. 1998; Smith and Todd 2001). One of the most commonly used quasi-experimental methods is propensity score matching (PSM), which selects project beneficiaries and nonbeneficiaries who are as similar as possible in terms of those observable characteristics expected to affect project participation and outcomes.[7] The difference in out-

---

[6] For example, a pastoralist in the state of Niger reported that he did not want to participate in Fadama II because similar projects in the past had failed.

[7] This method is referred to as quasi-experimental because it seeks to mimic the approach of experiments in identifying similar treatment and control groups. However, because the comparison groups identified by PSM are not selected by random assignment, they may differ in unobserved characteristics.

comes between the two matched groups can be interpreted as the impact of the project on the beneficiaries (Smith and Todd 2001). We used this method to estimate the ATT for impacts of Fadama II on household productive assets, incomes, and indicators of access to and impact of rural infrastructure.

The PSM method matches project beneficiaries with comparable nonbeneficiaries using a propensity score, which is the estimated probability of being included in the project. Only beneficiaries and nonbeneficiaries with comparable propensity scores are used to estimate the ATT. Those who do not have comparable propensity scores are dropped from the comparison groups. In our study, 1,728 of 3,758 observations matched. Therefore we used only the matched observations to analyze the impact of Fadama II.

Among the advantages of PSM over econometric regression methods is that it compares only comparable observations and does not rely on parametric assumptions to identify the impacts of projects. However, PSM is subject to the problem of selection on unobservables, meaning that the beneficiary and comparison groups may differ in unobservable characteristics, even though they are matched in terms of observable characteristics (Heckman et al. 1998). Econometric regression methods devised to address this problem suffer from the problems previously noted. As Heckman et al. (1998) further note, the bias resulting from comparing noncomparable observations can be much larger than that resulting from selection on unobservables, although this comparison may not be conclusively generalized.

In this study, we address the problem of selection on unobservables by combining PSM with the use of the double-difference estimator.[8] This estimator compares changes in outcome measures (the change from before to after the project) between project participants and nonparticipants, rather than simply comparing outcome levels at one point in time:

$$DD = (Y_{p1} - Y_{p0}) - (Y_{np1} - Y_{np0}), \tag{3}$$

where DD is the double-difference estimator; Y_{p0} and Y_{p1} are the outcomes of participants before and after project start, respectively; and Y_{np0} and Y_{np1} are the outcomes of nonparticipants before and after project start, respectively.

The advantage of the double-difference estimator is that it nets out the effects of any additive factors (whether observable or unobservable) that have fixed (time-invariant) impacts on the outcome indicator (such as the abilities of farmers or the inherent quality of natural resources) or that

[8] The double-difference method is also known as the difference-in-difference method (Duflo, Mullainathan, and Bertrand 2004).

reflect common trends affecting project participants and nonparticipants equally (such as changes in prices or weather) (Ravallion 2005). For example, if project participants and nonparticipants are different in their asset endowments (mostly observable) or in their abilities (mostly unobservable), and if those differences have an additive and fixed effect on outcomes during the period studied, such differences will have no confounding effect on the estimated ATT. Given that a large share of observations did not match, we sought to understand the impact of Fadama II on the unmatched beneficiaries. We did this evaluation by comparing the change in income and value of asset of beneficiaries and nonbeneficiaries, both of whom did not match; that is, we estimated *DD* using the straightforward equation (3), because these observations did not match. This comparison helps us to better understand the variation of the impact of the program across a wide range of beneficiaries.

In principle, the double-difference approach can be used to assess project impacts without using PSM and will produce unbiased estimates of impact as long as these assumptions hold. However, if the project has differential effects on people with different levels of wealth or other observable characteristics, the simple double-difference estimator will produce biased estimates if participant and nonparticipant households differ in those characteristics (Ravallion 2005). By combining PSM with the double-difference estimator, controls for differences in preproject observable characteristics can be established. A bias could still result from the heterogeneous or time-variant impacts of the unobservable differences between participants and nonparticipants. For example, communities and households that participated in Fadama I may have different responses to Fadama II than those that did not because of the cumulative effects of social capital developed under Fadama I, favorable or adverse experiences under Fadama I, or other factors.[9] Such shortcomings are unfortunately inherent in all nonexperimental methods of impact assessment (Duflo, Glennerster, and Kremer 2006). Although no solution to these potential problems is perfect, we believe the method we have used addresses these issues as well as possible in this case.

The standard errors estimated by the double-difference method may be inconsistent because of serial correlation or other causes of a lack of independence among the errors. In ordinary regression models, serial correlation can result from unobserved fixed effects, but by taking first differences, the double-difference method eliminates that source of serial correlation. However, serial correlation may still be a problem if more than two years of panel data are used (Duflo, Mullainathan, and Bertrand 2004). In our study,

---

[9] Unfortunately, we did not collect information on respondents' participation in Fadama I and thus could not try to test or control directly for such effects.

because we used only two periods (before and after project start), we are not concerned about serial correlation among multiple periods. Another reason for the possible dependence of the errors is clustering of the sample.

The propensity scores were computed using binary probit regression models. We estimated three probit models for three comparisons: (1) Fadama II beneficiaries compared with all nonbeneficiaries, (2) Fadama II beneficiaries compared with nonbeneficiaries in Fadama II communities, and (3) Fadama II beneficiaries compared with nonbeneficiaries outside Fadama II communities. The dependent variable in each model is a binary variable indicating whether the household was a beneficiary of the project.

The explanatory variables used in computing the propensity scores are those expected to jointly determine the probability to participate in the project and the outcome. We focused on the determinants of income and productive assets when selecting the independent variables for computing the PSM. We assumed that rural infrastructure should be included in productive assets. These variables are summarized in Table 4.2.

Consistent with the CDD approach, Fadama II supported economic groups only. Hence to better understand participation in Fadama II, we analyzed the determinants of membership in EIGs. This analysis adds more information to the PSM analysis, because the PSM model assessed the determinants of membership in Fadama II only, whereas the EIG analysis involved any economic group—even those that did not qualify or were not covered by Fadama II. We used the same covariates as those used for the PSM model (Table 4.2).

Elite capture is one of the potential problems occurring in CDD projects. If the program benefits accrue more to the well-off than to poor beneficiaries, income distribution will be more skewed, leading to increased inequity. We analyzed the impact of the program across asset terciles and agroecological zones. We divided the beneficiaries into three groups of poverty terciles using the value of productive assets prior to the program as an indicator of wealth.

To understand the impact of Fadama II on income distribution, we computed the Gini coefficient and the coefficient of variation for beneficiaries and nonbeneficiaries before and after the project.[10] We used both household consumption expenditure as well as income to measure the Gini coefficient. Each of these measures has its advantages and drawbacks. The main disadvantage of the consumption data is measurement error. Consumption expenditure was collected using the household survey, and farmers were asked

[10] The Gini coefficient is a measure of inequality, ranging from zero if income (or any other statistic) is equal across all members of a society to one if income (or any other statistic) belongs to or characterizes only one person in the society.

Table 4.2 Variables used to compute propensity scores, the probability that households participate in EIGs, and the expected trend in the effects of participation in Fadama II

Variable	Expected impact on participation in Fadama II / EIG	Why?	Expected trend in income and wealth	Why?
Gender of household head (female = 1, male = 0)	+	Fadama II has components targeting women's groups	-	Female-headed households are usually poorer than households headed by men
Household size	+	Larger families are often associated with poverty or other vulnerabilities that qualify them for Fadama II support	-	The larger the family, the poorer
Age of household head	+/-	Fadama II offers support for both the elderly and youth	+	Older household heads likely to be better off than young ones because of accumulation of wealth and experience over the life cycle
Level of education of household head (years of formal education)	+	Some projects require a certain level of education[a]	+	Education increases income opportunities, such as nonfarm activities
Area of rainfed land (ha)	+/-	Wealthier households more likely to join Fadama II because they are better able to pay the beneficiary contribution than less wealthy households; however, the project also supports the poor	+	Possession of more land enables households to invest more and get higher income and more productive assets

Agroecological zone (compared with humid forest zone)				
Moist savannah	?	Unknown	-	Humid forest zone is closer to major cities and has higher agroecological potential than the savannah zones
Dry savannah	?	Unknown	-	Same as above
Distance to nearest town before project started (km)	+	Requirement for bank account gives advantage to people living closer to roads and towns where banks always operate	+	Access to market increases income opportunities and reduces transaction costs
Distance to nearest all-weather road before project (km)	+	Same as above	+	Same as above
Value of productive assets before project (nairas)	+	Same as for land area	+	Same as for land area
Value of livestock before project (nairas)	+	Same as for land area	+	Same as for land area

Note: EIG, economic interest group.

[a]For example, Fadama User Groups qualifying for Fadama II support were required to have a bank account, a requirement that calls for a certain degree of education.

to estimate their overall consumption expenditure for two years, one year before the project and a year after project onset. The major consumption expenditure reported was for all household expenditures, including food, clothing, health, education, and transportation. Given the difficulty of recalling consumption expenditure over the long span of time, the data are of limited reliability.

Although the income data are significantly more reliable, the problem that arises here is that the presence of negative values for income, which is not uncommon, can render Gini coefficient values that are greater than 1, and such values are not subject to the common interpretation of a Gini coefficient (Chen, Tsaur, and Rhai 1982; Berrebi and Silber 1985; Stich 1996). In our case, income was negative for 32 percent of the surveyed households. Although the common meaning of the Gini coefficient is invalid for values greater than 1, an interpretation of an ordinal nature is still retained. That is, greater Gini coefficients are interpreted as indicating greater inequality, regardless of whether the measure is in the standard range. Therefore, we also present income inequality indicators and interpret them according to this convention. In addition, we also show inequality results using the income data after all negative income values have been set to zero. Of course, this practice introduces a strong downward bias on the inequality measures. The cutoff is problematic only if we expect the extent of downward bias on the Gini to differ between beneficiaries and nonbeneficiaries. Chapter 5 further discusses this problem. There we present results from all three welfare measures (consumption expenditure, and income and income with negative values normalized to zero).

Community infrastructure, demand-driven advisory services, and other Fadama II benefits are public and hard to limit to participants. Hence we expect spillover of Fadama II benefits to nonparticipants in Fadama II LGAs. Thus we treat the nonparticipants in Fadama II LGAs as a treated group and nonparticipants in non-Fadama II LGAs a control group. We estimated PSM to match the observations in the two groups and then estimated ATT as discussed earlier in the chapter. The ATT shows the magnitude of spillovers, because it measures the change in the outcome as a result of spillovers.

The results of the probit models are reported in Appendixes A and B. We found that Fadama II beneficiaries are more likely to be from female-headed and larger households than are nonbeneficiaries (both in and outside Fadama II LGAs). However, we observed that gender did not significantly affect membership in EIGs, suggesting that targeting of female household heads significantly increased their probability of participating in the program compared to male household heads. Compared with nonparticipants in Fadama II communities, Fadama II participants also tend to have older heads of household.

In contrast, participants tend to be younger and have more land but reside farther from an all-weather road compared with nonparticipants outside Fadama II communities. These results suggest that Fadama II targets vulnerable groups, such as female household heads, larger households, and people in more remote locations, although apparently the project also targets communities with relatively large farms. It does not select for other factors, such as education, ownership of productive assets or livestock, and agroecological zone. We also observe that younger people and those in remote areas were more likely to participate in EIGs. Education also increased the propensity to participate in EIGs.

These probit model results were used to compute the propensity scores that determined the PSM estimate of ATT. Several methods are possible for selecting matching observations (Smith and Todd 2001). We used the kernel matching method (using the normal density kernel), which uses a weighted average of neighbors (those observations within a given range in terms of the propensity score) of a particular observation to compute matching observations. Unlike the nearest-neighbor method, using a weighted average improves the efficiency of the estimator (Smith and Todd 2001). Observations outside the common range of propensity scores for both groups (meaning those lacking common support) were dropped from the analysis. This requirement of common support eliminated about half of the total number of observations, indicating that many of the observations from the various strata were not comparable.

Further testing of the comparability of the selected groups was done using a balancing test (Dehejia and Wahba 2002), which tests for statistically significant differences in the means of the explanatory variables used in the probit models between the matched groups of Fadama II participants and nonparticipants. In all cases, that test showed statistically insignificant differences in observable characteristics between the matched groups (but not between the unmatched samples), supporting the contention that PSM ensures the comparability of the comparison groups (at least in terms of observable characteristics).

We used bootstrapping to compute the standard errors of the estimated ATT, generating robust standard errors, because the matching procedure matched control households to treatment households "with replacement" (see Abadie and Imbens [2002, 2006] on the use of bootstrapping for inference in matching estimators).[11] Given that FUGs were managed by FCAs located

[11] Sampling with replacement means that the sampled observation is replaced such that with repeated sampling the probability of it being randomly picked for each new draw remains the same.

in LGAs, we expect some form of correlation among households in any given LGA. To account for this correlation, we estimated the bootstrapped standard errors of ATT with an option of clusters at the LGA level (Stata 2007). Given that we expect a strong correlation of the outcome in a given cluster (LGA), we expect that clustered standard errors will be larger than is the case for standard errors estimated without clustering. Hence statistical inferences from clustered standard errors are expected to be conservative.[12]

Using the matched samples, we also analyzed the impact of Fadama II on demand for advisory services. In that analysis, we compared the type and rate of adoption of production and postproduction technologies of Fadama II beneficiaries and nonbeneficiaries. We also asked the respondents using each technology whether they asked for that technology. We then compared the type of technologies demanded by Fadama II beneficiaries and nonbeneficiaries.

Because agriculture was among the major sectors supported by Fadama II and the adoption of improved production and marketing technologies is among the strategies that beneficiaries could use to increase their incomes, we analyzed the determinants of adoption decisions. To determine the impact of the program on adoption decisions, participation in Fadama II was included as one of the covariates. Because participation in Fadama II is an endogenous variable, conventional methods (such as fixed-effect methods with panel data) will produce biased estimates. We address this problem using a two-stage procedure, in which the estimated PSMs are used as weights in the regression model; the PSM weighting removes the bias stemming from any correlation between covariates and participation in Fadama II (Imbens and Wooldridge 2008). The two-stage weighted regression is specified as

$$\Delta Y_i = \beta_0 Y_0 + \beta_i \Delta X + \tau FII, \tag{4}$$

where Y_i is outcome i (income or value of assets), $i = 1, 2$; Y_0 is the initial value of the outcome of interest, X is the vector of covariates that determine outcome Y_i; and τ is a coefficient that measures the impact of FII. The vector X includes the same variables used for calculating PSM, because PSM is estimated using covariates that simultaneously affect both participation in the Fadama II and the outcomes of the program.

---

[12] Note that the estimation of standard errors using clusters affects only the standard errors and not the coefficients.

CHAPTER 5

Empirical Results

Major Sources of Income

Table 5.1 shows that crop production was a major source of income for all types of households before and after the start of Fadama II. The enterprise contributed more than 46 percent to the incomes of both beneficiaries and nonbeneficiaries before and after project initiation. The contribution of crop production to household income had increased one year after project start for all types of respondents, but the change was especially large (more than 10 percent) for Fadama II beneficiaries and those nonbeneficiaries outside Fadama II communities. For the Fadama II beneficiaries, the increased contribution of crops to household income could have resulted from the acquisition of productive assets that helped to add value (such as agroprocessing equipment) or increase productivity (such as irrigation). This result reflects the Fadama II focus on agriculture-based subprojects. Because Fadama II is a CDD project, the change also reflects the beneficiaries' demand for agricultural equipment and advisory services that led to increases in crop production. The factors contributing to the change in the contribution of crops to the incomes of households outside Fadama II communities remain unclear.

Nonfarm activities contributed the second-largest share of household income before and after project inception. The contribution of nonfarm activities to household income decreased for both the Fadama II beneficiaries and nonbeneficiaries living in communities not participating in Fadama II. This drop reflects an increase in the contribution of crops to income for those types of households. Fadama II supported both agricultural and nonfarm activities. The decrease in the contribution of nonfarm activities for Fadama II beneficiaries suggests that most chose to develop crop production and/or value addition for crops rather than participate in nonfarm activities. What is interesting is the low contribution of some activities that Fadama II supports. Beekeeping, hunting, gathering wild products, fish farming, and pastoral liveli-

Table 5.1 Sources of income for Fadama II beneficiaries and nonbeneficiaries

	Beneficiaries		Nonbeneficiaries in FII LGAs		Nonbeneficiaries outside FII LGAs	
Source of income	Before FII	After FII	Before FII	After FII	Before FII	After FII
Crops (nairas)	56,868	84,602	72,860	86,514	51,851	68,677
Share of total income (%)	46.600	56.800	53.800	56.500	46.900	60.200
Nonfarm (nairas)	34,428	48,724	26,566	46,367	46,416	51,805
Share of total income (%)	48.500	41.100	38.700	39.900	43.300	39.300
Livestock (nairas)	2,067	-7,754	4,444	3,504	3,931	2,219
Share of total income (%)	4.900	2.100	7.430	3.500	9.700	0.050
Other (nairas)	5,050	13,755	14,043	26,552	6,665	7,390
Share of total income (%)	0.001	0.001	0.130	0.140	0.190	0.540

Notes: "Before FII" indicates the year before Fadama II started (October 2004-September 2005). "After FII" indicates the year after the project started (October 2005-September 2006). FII, Fadama II; LGA, Local government authority.

hoods are among the activities the project encourages but that did not contribute significantly to the global average of household incomes reported in Table 5.1. However, for households that heavily depend on these enterprises (for example, in pastoral communities in northern Nigeria), their contribution to household income is large. Because the project is a CDD, the limited contribution of those activities suggests that few beneficiaries demanded them.

Impact of Fadama II on PAA

PAA is the second-largest investment undertaken by Fadama II, after rural infrastructure investments (World Bank 2003). Because Fadama II supported PAA by FUGs rather than by individual households, we divided the productive assets into those owned by individual farmers and those owned jointly by EIGs. It was not easy to determine the share of value that each member of a group held in jointly owned productive assets. The intensity of use of the productive assets also differed across households within groups. For example, members of an EIG owning a borehole for watering animals used the equipment not according to how much they contributed but according to their needs as determined by the number of animals they owned. Our data collection focused on the household-level assets and did not capture the group-level management of productive assets. Figures 5.1 and 5.2 show that Fadama II had a large and statistically significant impact on the value of pro-

Figure 5.1 Change in the value of group-owned productive assets resulting from participation in Fadama II

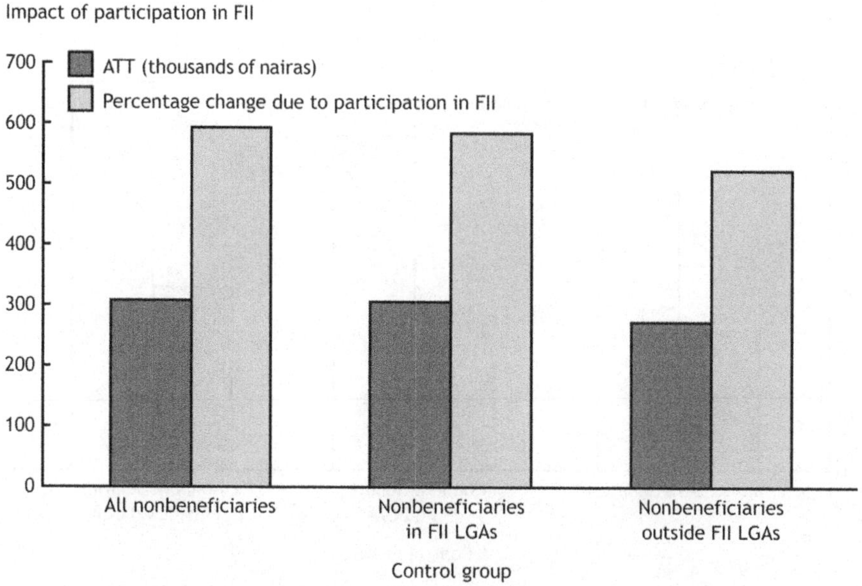

Notes: ATT and the corresponding percentage change refer to the change in productive assets resulting from participation in Fadama II compared with the corresponding group of nonbeneficiaries. Thus these values should not be interpreted as referring to the change in productive assets of the corresponding control group of nonbeneficiaries. ATT, average effect of the treatment on the treated; FII, Fadama II; LGA, local government authority.

ductive assets owned by groups and individuals benefiting from the project compared with nonbeneficiaries.

In all comparisons reported in Table 5.2, Fadama II beneficiaries saw the value of group-owned productive assets increase significantly (at a statistical significance level of $p = 0.01$) across all agroecological zones (except the dry savannah, where the increase was significant only at $p = 0.10$ for the unclustered standard-error case), all asset terciles, and both genders. The humid forest zone recorded the largest and significant increase in the absolute value of group-owned assets, whereas the dry savannah zone—where the level of poverty is most severe—reported the lowest (and nonsignificant) increase in percentage and absolute value. The same pattern is observed in the unmatched sample results (Table 5.3).

The poorest tercile of beneficiaries (in terms of value of assets owned before project start) experienced the largest increase in group-owned pro-

Figure 5.2 Change in the value of individually owned assets resulting from participation in Fadama II

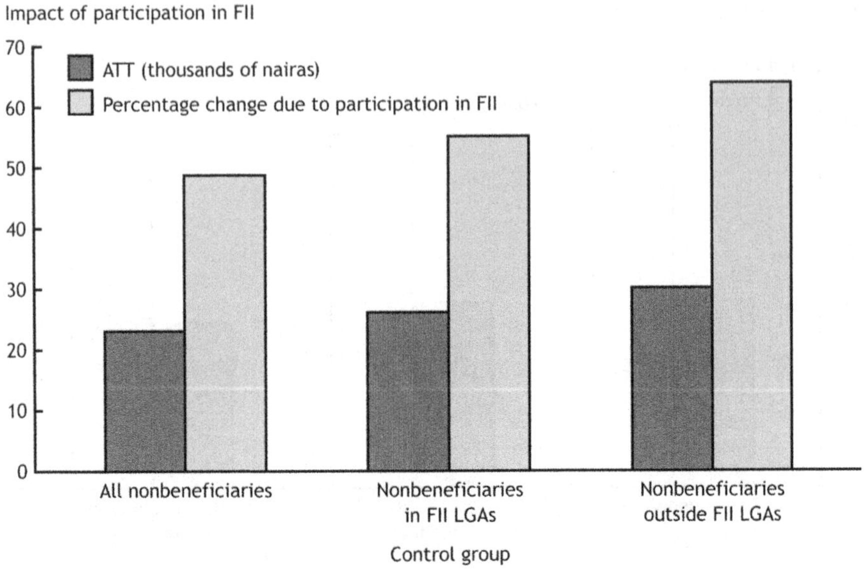

Note: ATT, average effect of the treatment on the treated; FII, Fadama II; LGA, local government authority.

ductive assets (both in terms of absolute value and percentage): an average increase of 91,780 percent (from only 482 to 470,865 nairas).[1] The unmatched sample results show a similar pattern—the value of the group assets increased more significantly both in absolute value and percentage (Table 5.3). The reason for this massive increase is that ownership of group productive assets was relatively small for those beneficiaries before the project.[2] The large increase in the value of jointly owned productive assets includes the value of the cash transfer (70 percent of the total productive-asset value) from the project to the beneficiaries.

[1] This increment is not a simple difference between the before and after values. Rather, it is an increase that takes into account the changes of the control group, that is, ATT / value of assets of beneficiary before the project. These values are all in real nairas (deflated to 2003 value).

[2] However, the preproject level of group assets was significantly larger for Fadama II beneficiaries than for nonbeneficiaries. This difference might result from a greater tendency of Fadama II beneficiaries to have participated in group activities before the project compared to nonbeneficiaries. But it might reflect a reporting error concerning when group assets were acquired by Fadama II beneficiaries (that is, some Fadama II respondents may have mistakenly reported some of the group assets that they acquired under Fadama II as group assets owned before project inception). If the second case is true, then the impacts of Fadama II on the acquisition of group assets have been underestimated.

The most common FUG productive assets acquired were water and irrigation equipment, which 118 of 489 Fadama II households (24 percent) obtained (Table 5.4). The value of FUG water and irrigation equipment increased by 2,771 percent (from 47,475 nairas before the project to 1,362,937 nairas by September 2006), highlighting the large impact that Fadama II had on the value of productive assets. Furthermore, individually owned water and irrigation assets more than doubled in value over the same period. Total values of processing equipment, livestock, and building structures owned by FUGs more than doubled. The large increases for individual productive-asset types add up to a large rise in the total value of productive assets, especially for beneficiaries in the poorest asset tercile, who had few productive assets before the project.

The percentage increase in value of group-owned productive assets in the upper asset tercile that was due to Fadama II participation was only 63 percent—the smallest—but the absolute value (of ATT) was the second largest. The value of productive assets owned by women's EIGs participating in the project also increased significantly compared with that belonging to women's groups not participating in the project. These results demonstrate that the PAA component succeeded in its efforts to target poor and vulnerable groups.

Compared with all nonbeneficiaries and with nonbeneficiaries in and outside Fadama II communities, project beneficiaries experienced greater increases in the value of individually owned productive assets. The impact of Fadama II on these assets was not significant across all zones for the clustered standard errors. Comparisons between male beneficiaries and nonbeneficiaries showed significantly greater percentage and absolute value increases in the value of private productive assets for beneficiaries. Comparison of the ATT of individually owned assets for female-headed household beneficiaries and nonbeneficiaries showed no significant difference for both clustered and unclustered standard errors, suggesting that the impact of the program on female heads of households did not trickle down to private assets, which the program does not support (NFDO 2006). This result is to be expected, given that the poor cannot simultaneously afford to pay their share of the 30 percent matching funds to buy the group-owned assets and at the same time buy private assets.

Surprisingly, the value of individually owned productive assets of beneficiaries and nonbeneficiaries decreased significantly for the humid zone, for all women's groups, and for all groups in asset terciles 2 and 3 (the wealthier terciles) (see Table 5.3). One possible reason for this trend could be the lack of credit that forces potential Fadama II beneficiaries to liquidate their private assets. This explanation is supported by the corresponding increase in group-owned assets for the beneficiaries in most groups studied. But as shown

Table 5.2 Value of productive assets before and after Fadama II across agroecological zones, gender, and asset terciles (matched sample)

Characteristic / treatment type	Value of group-owned assets				Value of individually owned assets			
	Before FII[a] (nairas)	After FII[a] (nairas)	ATT[b] (nairas)	Change[b] (%)	Before FII[a] (nairas)	After FII[a] (nairas)	ATT[b] (nairas)	Change[b] (%)
Agroecological zone								
Humid forest zone								
FII beneficiaries (*n* =176)	83,467 (408,783)	619,889 (1,287,487)	577,722***/††† (155,678)	692	72,634 (160,061)	86,552 (206,851)	7,628 (14,340)	10.5
All nonbeneficiaries (*n* = 282)	7,724 (60,698)	3,087 (61,750)			75,986 (148,462)	74,963 (130,112)		
Moist savannah zone								
FII beneficiaries (*n* = 176)	9,810 (34,924)	397,858 (631,549)	446,230***/† (103,044)	4,549	74,640 (133,834)	103,899 (193,368)	61,342*/NS (36,419)	82.2
All nonbeneficiaries (*n* = 282)	5,957 (86,649)	11,943 (82,400)			47,909 (96,834)	43,831 (122,102)		
Dry savannah zone								
FII beneficiaries (*n* = 176)	46,074 (168,351)	68,383 (353,121)	44,307*/NS (25,334)	96	37,060 (41,595)	43,579 (51,842)	19,965***/NS (5,644)	53.9
All nonbeneficiaries (*n* = 282)	4,248 (40,717)	963 (17,359)			40,372 (46,050)	36,716 (69,737)		
Gender of household head								
Female								
FII beneficiaries (*n* = 176)	28,651 (229,778)	505,381 (1,125,385)	448,254***/†† (124,701)	1,565	51,572 (125,764)	74,202 (160,235)	16,701 (22,733)	32.4
All nonbeneficiaries (*n* = 282)	6,826 (65,012)	6,526 (82,256)			63,531 (151,330)	55,318 (158,418)		

Male								
FII beneficiaries (n = 176)	65,774	260,596	217,443***/†	331	55,064	62,256	41,504**/††	75.4
	(296,916)	(738,377)	(49,500)		(75,064)	(137,669)	(19,286)	
All nonbeneficiaries (n = 282)	5,715	4,181			49,812	53,054		
	(62,375)	(48,739)			(67,000)	(97,857)		
Asset tercile								
Tercile 1 (poorest)								
FII beneficiaries (n = 176)	482	470,865	442,471>***/††	91,780	5,228	52,936	6,682	127.8
	(2,370)	(1,072,014)	(76,891)		(4,711)	(121,364)	(6,922)	
All nonbeneficiaries (n = 282)	123	3,687			7,642	47,468		
	(1,225)	(46,906)			(18,753)	(111,176)		
Tercile 2								
FII beneficiaries (n = 176)	3,573	213,483	104,922***/††	2,937	44,546	44,699	27,844***/†	62.5
	(13,619)	(629,129)	(39,112)		(23,112)	(45,656)	(6,383)	
All nonbeneficiaries (n = 282)	1,460	1,924			51,048	42,397		
	(8,460)	(18,802)			(24,780)	(61,513)		
Tercile 3 (wealthiest)								
FII beneficiaries (n = 176)	236,793	130,155	149,799***/†	63	99,577	124,724	80,174***/†††	80.5
	(554,079)	(436,402)	(60,733)		(52,972)	(237,442)	(21,579)	
All nonbeneficiaries (n = 282)	31,447	11,755			114,505	95,846		
	(146,968)	(104,818)			(49,934)	(148,597)		

Notes: Numbers in parentheses are standard deviations of the corresponding mean (without clustering). *, **, and *** indicate significance at the 10, 5, and 1 percent levels, respectively, without clustering; †, ††, and ††† indicate significance at the 10, 5, and 1 percent levels, respectively, with clustering. NS means that the clustered standard error (SE) is not significant but the unclustered SE is significant at least at $p = 0.10$. Coefficients without * or † are not significant for both clustered and unclustered SEs. ATT, average effect of treatment on the treated; FII, Fadama II.

[a]"Before FII" indicates the year before Fadama II started (October 2004-September 2005). "After FII" indicates the year after the project started (October 2005-September 2006).

[b]ATT is computed as $[E(Y_1|p = 1) - E(Y_0|p = 0)] - [E(Y_0|p = 1) - E(Y_0|p = 0)]$. ATT and the corresponding Change column refer to the change in productive assets resulting from participation in Fadama II compared with the corresponding group of nonbeneficiaries. Thus these values should not be interpreted as referring to the change in the productive assets of the corresponding control group of nonbeneficiaries.

Table 5.3 Value of productive assets before and after Fadama II across agroecological zones, genders, and asset terciles (unmatched sample)

	Value of group-owned assets				Value of individually owned assets			
Characteristic / treatment type	Before FII[a] (nairas)	After FII[a] (nairas)	Double difference[b] (nairas)	Change due to participation[c] (%)	Before FII[a] (nairas)	After FII[a] (nairas)	Double difference[b] (nairas)	Change due to participation[c] (%)
FII beneficiaries (n = 660)	24,236	155,841	135,671***	559	24,554	16,331	21,706***	88
All nonbeneficiaries (n = 1,388)	8,749	4,682			35,779	5,850		
Agroecological zone								
Humid forest zone								
FII beneficiaries (n = 100)	7,791	263,640	254,703***	3,269	13,629	10,796	10,515	77
All nonbeneficiaries (n = 185)	671	1,817			20,629	7,296		
Moist savannah zone								
FII beneficiaries (n = 200)	26,639	283,165	246,386***	925	13,958	19,980	27,400***	196
All nonbeneficiaries (n = 399)	452	10,591			24,312	2,933		
Dry savannah zone								
FII beneficiaries (n = 360)	27,471	54,526	39,690***	144	33,605	15,826	20,468***	61
All nonbeneficiaries (n = 804)	14,949	2,313			45,258	7,012		

Gender of household head								
Female								
FII beneficiaries (*n* = 257)	8,977	153,040	144,942***	1,614	13,446	8,769	9,194**	68
All nonbeneficiaries (*n* = 388)	4,625	3,746			15,818	1,947		
Male								
FII beneficiaries (*n* = 403)	34,029	157,638	128,989***	379	31,736	21,221	26,105***	82
All nonbeneficiaries (*n* = 987)	10,494	5,114			44,088	7,467		
Asset tercile								
Tercile 1 (poorest)								
FII beneficiaries (*n* = 454)	519	176,380	173,187***	33,369	5,177	13,879	11,060***	213
All nonbeneficiaries (*n* = 877)	62	2,735			2,890	5,106		
Tercile 2								
FII beneficiaries (*n* = 109)	5,904	125,679	110,188***	1,866	47,720	21,570	11,027**	23
All nonbeneficiaries (*n* = 305)	1,193	10,780			45,720	8,917		
Tercile3 (wealthiest)								
FII beneficiaries (*n* = 95)	160,809	94,713	5,881	4	169,097	22,090	82,441***	48
All nonbeneficiaries (*n* = 197)	64,460	4,245			104,555	4,191		

Notes: ** and *** indicate significance at the 5 and 1 percent levels, respectively. FII, Fadama II.
[a]"Before FII" indicates the year before Fadama II started (October 2004–September 2005). "After FII" indicates the year after the project started (October 2005–September 2006).
[b]Double difference (DD) is calculated as $(Y_{p1} - Y_{p0}) - (Y_{np1} - Y_{np0})$. The variables are defined in the text.
[c]Change due to participation in project is calculated as $(DD/Y_{p0}) \times 100$; Y_{p0} is defined in the text.

Table 5.4 Value of productive assets for Fadama II beneficiaries (matched sample) (nairas)

	Value of productive asset			
	Group-owned		Individually owned	
Type of asset	Before FII[a]	After FII[a]	Before FII[a]	After FII[a]
Production equipment	71,944 (148,483) (*n* = 18)	158,888 (156,116) (*n* = 18)	38,335 (74,809) (*n* = 65)	52,856 (70,038) (*n* = 65)
Transport equipment	176,882 (122,897) (*n* = 17)	194,529 (117,323) (*n* = 17)	66,513 (95,992) (*n* = 127)	86,485 (115,898) (*n* = 127)
Processing equipment	165,149 (740,261) (*n* = 69)	527,011 (793,466) (*n* = 69)	49,440 (87,664) (*n* = 69)	59,512 (84,749) (*n* = 69)
Fishing equipment	43,422 (53,878) (*n* = 27)	147,674 (167,484) (*n* = 27)	111,187 (326,758) (*n* = 41)	91,589 (174,255) (*n* = 41)
Water and irrigation equipment	47,475 (205,301) (*n* = 118)	1,362,937 (1,440,951) (*n* = 118)	17,000 (28,967) (*n* = 74)	63,331 (124,446) (*n* = 74)
Livestock equipment	38,482 (113,752) (*n* = 31)	447,900 (492,751) (*n* = 31)	16,964 (34,555) (*n* = 49)	41,385 (97,515) (*n* = 49)
Building structures	139,903 (624,995) (*n* = 31)	512,419 (1,018,658) (*n* = 31)	92,504 (203,157) (*n* = 50)	119,024 (232,709) (*n* = 50)

Notes: Number in parentheses is the standard deviation of the corresponding mean. Production equipment = ox plow, oxen, tractor; transport equipment = bicycle, wheelbarrow, pickup truck, motorcycle, other means of transport; processing equipment = honey-processing equipment, milling machine, refrigerator, other processing equipment; fishing equipment = fishing gear, canoe, fishing boat engine; water and irrigation equipment = water pump, borehole, tube well; livestock equipment = cattle pen, cattle trough; building equipment = storage, fishpond. FII, Fadama II.

[a]"Before FII" indicates the year before Fadama II started (October 2004-September 2005). "After FII" indicates the year after the project started (October 2005-September 2006).

above, there was generally an increase in the value of individually owned assets for most groups, even though in some cases, such increases were not significant and/or declined slightly. The marked differences of the matched and unmatching groups are likely due to the factors analyzed in the propensity score regression.

Overall, the increase in the value of productive assets was generally less for individually owned productive assets than for those owned by EIGs. Even though Fadama II did not support individuals in purchasing productive assets, FUG members—especially male-headed households—were able to acquire such productive assets through their groups. The individual acquiring the private productive asset would pay the entire beneficiary contribution in the name of the FUG. Fadama II did not interfere with the private ownership of productive assets, which could explain the significant increase in the value of such assets for beneficiaries. Another possible explanation is that FUG members were required to buy complementary inputs to support the jointly owned productive assets. For example, FUG members owning irrigation equipment may have needed to buy pesticide sprayers to grow irrigated vegetables. The statistically insignificant impact of project participation on private productive assets for beneficiaries in the poorest asset tercile and in female-headed households suggests that the poor and vulnerable were not able to finance both types of assets. The results suggest that targeting was not effective in increasing the individually owned assets, which are important for increasing the efficiency of group-owned productive assets.

However, the estimated magnitude of the mean impacts for these groups was positive and large (128 percent increase for the poorest asset tercile and 32 percent for female-headed households), even though these estimates were not statistically significant. Therefore, the statistical insignificance of the estimates does not prove that the impacts were nonexistent; rather, it indicates that the variances of subsample impacts were too large to measure with the sample size we had. For example, the unmatched sample results show that participation in Fadama II significantly increased the value of individually owned assets for the poorest asset tercile (see Table 5.3). The differences in the impact of matched and unmatched samples could be due to significant differences in the initial values of individually owned assets. As shown in Appendix C, the values of private productive assets for the unmatched beneficiaries and nonbeneficiaries before Fadama II were significantly lower than those for the matched beneficiaries and nonbeneficiaries. The small initial value of these assets could explain the significant impact on poor beneficiaries. However, the absolute increase of the productive assets in the richest tercile was the greatest, suggesting that the wealthy benefited more than their poor compatriots, even though their percentage increase was smaller.

How are these productive assets managed, and how are their benefits shared among group members? These interesting questions require further study of the efficiency of collective ownership of productive assets and how the poor among FUG members benefit from such assets. Issues to investigate include the economic viability, maintenance, management, and operational efficiency of these assets. Among the benefits of studying jointly owned productive assets are a greater understanding of the returns to productive assets and how they affect the productivity of labor and other resources, and increased knowledge of methods for targeting poor and vulnerable groups and how they benefit from productive assets. Our study was conducted at the household level and did not capture these aspects for jointly owned productive assets. However, we did investigate the impacts of participation in Fadama II on household incomes, which reflects the effects of acquisition and use of both group and private productive assets, as well as other components of the project (such as the effects of rural infrastructure investments and agricultural advisory services).

Another interesting question to explore is the sustainability of the Fadama II success story beyond the project period and how it can be replicated in communities that did not benefit from the project. The major constraint faced by poor households is their ability to finance the acquisition of high-value assets without some form of support from projects or credit services. Fadama II did not involve credit service providers because of the high interest they charge and their limited availability. Thus alternative sources of credit were used by the 14 percent of beneficiaries who had access to credit services (Table 5.5). Relatives, social clubs, and friends were reported to be the major sources of credit for Fadama beneficiaries as well as for non-beneficiaries in and outside Fadama II communities. This finding underscores the limited options of poor beneficiaries to pay their 30 percent contribution to productive assets.

It is not clear how the poor were able to pay their contributions and if they were able to manage assets efficiently.[3] Those who could not otherwise secure the necessary funds may have used financing through wealthier friends or relatives (see Table 5.5). For example, an eligible but poor beneficiary could have entered into a rental agreement whereby an ineligible rich person paid the beneficiary's contribution and then asked the beneficiary to pay a premium for a specified period, or to share use of the productive asset or part of the returns. In some cases, an ineligible person could own the productive asset after paying the contribution of all beneficiaries and then rent the productive asset back to the beneficiaries. For example, a woman in one FUG reported that she entered into a rental agreement with a wealthy man who paid her beneficiary contribution for a milling machine. Such arrangements could affect the targeting of the poorest.

The World Bank supervision mission of February 2007 noted that most of the subprojects for women and the vulnerable had not been implemented, because these groups could not pay their contributions (World Bank 2007a,

[3] It is still too early to tell how FUGs managed and benefited from their productive assets. However, the MTR concluded that the capacity to manage some productive assets was low and there was still need for building the capacity of FUGs to manage their assets efficiently (World Bank 2007a).

Table 5.5 Access to credit, 2005-06

Type of access	FII beneficiaries (*n* = 621)	Nonbeneficiaries: In FII LGAs (*n* = 568)	Nonbeneficiaries: Outside FII LGAs (*n* = 539)	Total	Test[a]
Has access to credit (share of households)	0.176 (0.381)	0.089 (0.286)	0.141 (0.348)	0.137 (0.343)	b,c
Source of credit (proportion of households with access)					
Banks	0.073 (0.262)	0.118 (0.325)	0.118 (0.325)	0.097 (0.297)	
Relatives, social clubs, and friends	0.220 (0.416)	0.250 (0.437)	0.289 (0.457)	0.249 (0.433)	
Cooperatives	0.186 (0.391)	0.308 (0.466)	0.179 (0.386)	0.210 (0.408)	
Farmer associations	0.046 (0.210)	0.020 (0.140)	0.013 (0.115)	0.030 (0.170)	
NACRDB	0.130 (0.338)	0.115 (0.323)	0.113 (0.318)	0.121 (0.327)	
Local government	0.018 (0.135)	0.059 (0.238)	0.013 (0.115)	0.025 (0.158)	
Nongovernmental organizations	0.037 (0.189)	0.000 (0.000)	0.013 (0.115)	0.021 (0.144)	
State government	0.037 (0.189)	0.039 (0.196)	0.039 (0.196)	0.038 (0.192)	
Fadama II	0.109 (0.313)	0.020 (0.140)	0.000 (0.000)	0.055 (0.228)	b,d
Other	0.046 (0.210)	0.000 (0.000)	0.039 (0.196)	0.034 (0.181)	

Notes: Numbers in parentheses are standard deviations. FII, Fadama II; LGA, local government authority; NACRDB, Nigerian Agricultural Cooperative and Rural Development Bank.

[a]Empty cells imply paired comparison of any two groups in the corresponding columns that are not statistically different at the 5 percent level.

[b]Difference between FII beneficiaries and nonbeneficiaries living in the same LGA is significant at the 5 percent level.

[c]Difference between nonbeneficiaries living in and those living outside FII LGAs is significant at the 5 percent level.

[d]Difference between FII beneficiaries and nonbeneficiaries living outside FII LGAs is significant at the 5 percent level.

2007b). The mission also noted that most of the processing equipment acquired by women was operated by hired hands who benefited more than the project beneficiaries. The Bank thus recommended that the beneficiary contribution for women and the vulnerable be reduced to 10 percent. Initially the project set the contribution of beneficiaries of the PAA component to 40 percent of the value of the productive asset (NFDO 2006), but reduced it to 30 percent because of overwhelming evidence of the failure of the poor to pay their share. Even the 30 percent contribution might be high for expensive productive assets and force FUG members who are unable to pay their contribution to turn to more wealthy individuals for credit support or rental arrangements. Planners for the next phase of the project (Fadama III) need to consider the use of sustainable financing for targeted groups—for example, through microfinancing institutions. Existing local rotating savings and credit schemes, such as *esusu, dashi,* and *adashi,* could help to increase credit access (Bascom 1952; Okonjo 1979; Bouman 1995).

Impact of Fadama II on Household Income

Figure 5.3 shows that the average annual household income after Fadama II started (2005-06) for all types of households ranged from 43,298 to 108,625 nairas (real value in 2003). The lower limit is above the average rural household income of 42,644 nairas reported by the 2003-04 living standard survey (FOS 2004) but of the same order of magnitude. On average, the real incomes of Fadama II beneficiaries increased 58.5 percent as a result of participation in the project, based on the PSM and double-difference estimation (ATT); that increase is well above the Fadama II target of 20 percent for half of the beneficiaries after six years of operation.

Results of the unmatched sample showed that the incomes of beneficiaries increased by 38 percent. In contrast, average real incomes of all nonbeneficiaries increased by only 15.5 percent and by even less for nonbeneficiaries outside Fadama II communities (12.7 percent).[4] The mean increase in income for beneficiaries was significantly different from that for nonbeneficiaries at $p = 0.05$. Considering the income of beneficiaries before and after the project (without controlling for other reasons for income to change), about 42 percent of beneficiaries increased their incomes by at least 20 percent in the first year of Fadama II (Table 5.6). In contrast, the share of nonbeneficiaries who increased their incomes by at least 20 percent was only 34 percent. Although this percentage includes the effects of other factors that influence income changes over time, it is clear that Fadama II achieved considerable success in its first year of operation.

It is likely that the impact of the project on incomes will be larger in the future because of the delayed effects of investments in productive assets, infrastructure, and other project investments. Even without longer term lags, the impacts on incomes in 2005-06 could be expected to be less than

---

[4] The percentage changes of the nonbeneficiaries before and after project inception are not reported in the table but are calculated using the following simple formula (symbols are as defined in equation (3)): $[(Y_{np1} - Y_{np0})/Y_{np0}] \times 100$.

Figure 5.3 Household income one year before and one year after Fadama II started

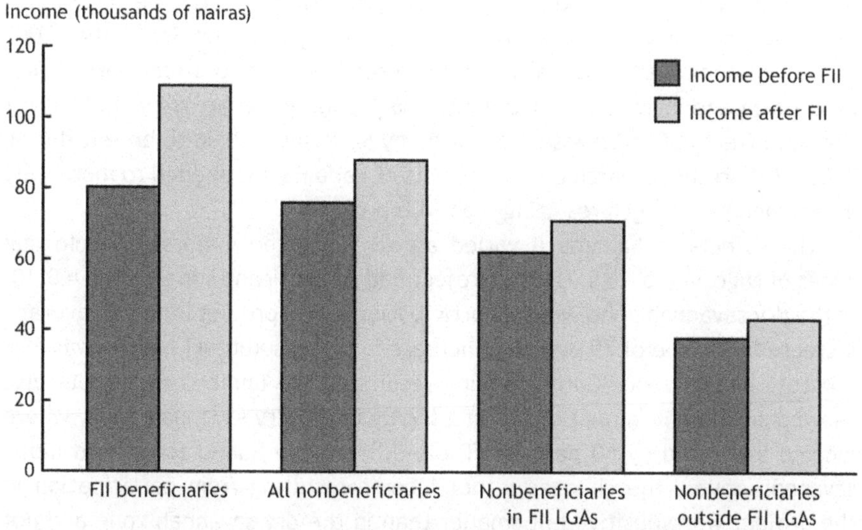

Notes: "Before FII" indicates the year before Fadama II started (October 2004-September 2005). "After FII" indicates the year after the project started (October 2005-September 2006). FII, Fadama II; LGA, local government authority.

Table 5.6 Achievement of target increase in income among Fadama II beneficiaries

	Change in real income before and after FII (%)[a]		
Treatment type	**≤20**	**20-50**	**≥50**
FII beneficiaries (*n* = 1,281)	57.9	11.9	30.1
All nonbeneficiaries (*n* = 1,240)	66.2	12.6	21.1
Nonbeneficiaries in FII LGAs (*n* = 1,240)	65.1	15.5	19.4
Nonbeneficiaries outside FII LGAs (*n* = 1,229)	67.5	9.2	23.2

Note: FII, Fadama II; LGA, local government authority.
[a]"Before FII" indicates the year before Fadama II started (October 2004-September 2005). "After FII" indicates the year after the project started (October 2005-September 2006).

proportionate to the increase in productive assets from September 2005 (at the beginning of project implementation) to September 2006 (the date for measuring changes in productive assets after project start), because many of the investments in productive assets occurring between September 2005 and September 2006 may not have come soon enough to affect agricultural production and income during the 2005-06 production year. We would expect the full effects of those assets acquired by September 2006 to be felt during 2006-07. Further research on the impacts of Fadama II is needed to more fully assess income changes resulting from the project.

The effects of Fadama II varied across the three major agroecological zones of Nigeria (Table 5.7). The project had a significant impact (at $p = 0.10$) in the dry savannah zone, where participation in the project led to an average increase in income of 79 percent. Increase in the absolute ATT value was also greatest in this zone. Corresponding results of the unmatched sample also showed significant impact of Fadama II only in the dry savannah zone, where income increased by 60 percent (Table 5.8). In the humid forest and moist savannah zones, the changes in net income resulting from participation in the project were positive but smaller than in the dry savannah zone and not statistically significant. The large net increase in income in the dry savannah zone, where limited rainfall is a major problem, could be explained by

Table 5.7 Impact of Fadama II on household income across agroecological zones, gender, and asset terciles (matched sample)

	Net real annual household income (nairas)			Net change due to participation[c]
Characteristic / treatment type	Before FII[a]	After FII[a]	ATT[b]	(%)
Agroecological zone				
Humid forest zone				
FII beneficiaries (n = 176)	87,431 (292,102)	112,626 (299,102)	14,963	17.1
All nonbeneficiaries (n = 282)	12,307 (257,170)	31,343 (276,530)		
Moist savannah zone				
FII beneficiaries (n = 118)	70,578 (203,342)	74,295 (280,596)	33,522	47.5
All nonbeneficiaries (n = 251)	96,498 (258,137)	77,384 (271,796)		
Dry savannah zone				
FII beneficiaries (n = 205)	79,113 (255,967)	124,458 (225,341)	62,664*/††	79.2
All nonbeneficiaries (n = 335)	106,066 (255,201)	142,708 (254,173)		

Table 5.7 Continued

Characteristic / treatment type	Net real annual household income (Nairas) Before FII[a]	After FII[a]	ATT[b]	Net change due to participation[c] (%)
Gender of household head				
FII beneficiaries				
Male (n = 311)	83,691 (280,998)	107,454 (282,103)	-749	-0.9
Female (n = 198)	74,284 (217,805)	110,454 (239,427)		
Female				
FII beneficiaries (n = 198)	74,326 (217,819)	110,383 (239,400)	51,303**/NS	69.1
All nonbeneficiaries (n = 178)	35,414 (210,009)	48,346 (219,474)		
Male				
FII beneficiaries (n = 674)	83,701 (281,080)	107,495 (282,132)	84,825***/††	101.3
All nonbeneficiaries (n = 267)	86,261 (269,010)	98,249 (281,306)		
Asset tercile				
Tercile 1 (poorest)				
FII beneficiaries (n = 293)	70,851 (154,438)	82,745 (153,922)	31,776	44.9
All nonbeneficiaries (n = 505)	76,831 (153,000)	77,511 (153,998)		
Tercile 2				
FII beneficiaries (n = 93)	93,847 (161,254)	119,013 (175,283)	94,750**/††	101.0
All nonbeneficiaries (n = 191)	74,705 (163,651)	104,994 (180,714)		
Tercile 3 (wealthiest)				
FII beneficiaries (n = 96)	122,074 (239,037)	154,892 (267,235)	1,177	1.0
All nonbeneficiaries (n = 139)	126,474 (207,494)	128,269 (223,225)		

Notes: Numbers in parentheses are standard deviations of the corresponding mean (without clustering). *, **, and *** indicate significance at the 10, 5, and 1 percent levels, respectively, without clustering; †† indicates significance at the 5 percent level with clustering. Coefficients without * or † are not significant for either clustered or unclustered standard errors. ATT, average effect of treatment on the treated; FII, Fadama II; NS, not significant.

[a]"Before FII" indicates the year before Fadama II started (October 2004-September 2005). "After FII" indicates the year after the project started (October 2005-September 2006).

[b]ATT is computed as $[E(Y_1|p = 1) - E(Y_0|p = 0)] - [E(Y_0|p = 1) - E(Y_0|p = 0)]$. See equation (1) in the text for definitions of the variables. ATT and the corresponding Change column refer to the change in income resulting from participation in Fadama II compared with the corresponding group of nonbeneficiaries. Thus they should not be interpreted as referring to the change in the income of the corresponding control group of nonbeneficiaries.

[c]Net change due to participation in Fadama II is calculated as $(ATT/Y_{p0}) \times 100$. Y_{p0} is defined in the text.

Table 5.8 Impact of Fadama II on household income across agroecological zones, gender, and asset terciles (unmatched sample)

Characteristic / treatment type	Household income			
	Before FII[a] (nairas)	After FII[a] (nairas)	Double difference[b] (nairas)	Change due to participation[c] (%)
FII beneficiaries (n = 660)	74,869	101,469	28,715**	38
All noneneficiaries (n = 1,388)	85,942	108,215		
Agroecological zone				
Humid forest zone				
FII beneficiaries(n = 100)	45,190	38,461	-32,957	-73
All nonbeneficiaries (n = 185)	35,562	44,881		
Moist savannah zone				
FII beneficiaries (n = 200)	75,040	78,430	23,212	31
All nonbeneficiaries (n = 399)	97,852	106,693		
Dry savannah zone				
FII beneficiaries (n = 360)	83,764	133,654	50,408***	60
All nonbeneficiaries (n = 804)	92,304	125,507		
Gender of household head				
Female				
FII beneficiaries (n = 257)	61,093	88,615	12,919	21
All nonbeneficiaries (n = 388)	50,716	67,455		
Male				
FII beneficiaries (n = 403)	84,539	110,950	40,255**	48
All nonbeneficiaries (n = 987)	102,064	126,886		
Asset tercile				
Tercile 1 (poorest)				
FII beneficiaries (n = 454)	66,384	89,535	29,664**	45
All nonbeneficiaries (n = 877)	78,823	105,937		
Tercile 2				
FII beneficiaries (n = 109)	100,630	117,934	27,744	28
All nonbeneficiaries (n = 305)	105,242	119,937		
Tercile 3 (wealthiest)				
FII beneficiaries (n = 95)	81,323	135,746	23,517	29
All nonbeneficiaries (n = 197)	95,346	105,431		

Notes: Numbers in parentheses are standard deviations of the corresponding mean (without clustering). ** and *** indicate significance at the 5 and 1 percent levels, respectively. FII, Fadama II.

[a]"Before FII" indicates the year before Fadama II started (October 2004-September 2005). "After FII" indicates the year after the project started (October 2005-September 2006).

[b]Double difference (DD) is calculated as $(Y_{p1} - Y_{p0}) - (Y_{np1} - Y_{np0})$. The variables are defined in the text.

[c]Change due to participation in project is calculated as $(DD/Y_{p0}) \times 100$; Yp_0 is defined in the text.

the acquisition of irrigation facilities and water equipment, which address a major production constraint in that zone.

A comparison of male- versus female-headed beneficiary households showed no significant difference in incomes before or after the project. Income changes for female-headed beneficiary households were significantly greater than those for female-headed nonbeneficiary households for the unclustered standard-error case but were not significant for the clustered standard-error case.

Similarly, in the unmatched sample, the change in incomes of female-headed beneficiary (compared to nonbeneficiary) households was not significant at 10 percent (Table 5.8). These results suggest that Fadama II had little impact on the short-term incomes of vulnerable beneficiaries. The results are consistent with those we observed for individually owned productive assets. We found that the project significantly increased income for male-headed beneficiary (relative to nonbeneficiary) households, with a higher estimated percentage and absolute value of ATT for male- than for female-headed households. We found comparable results for the unmatched sample (see Table 5.8). The results suggest male-headed households experienced a much larger short-term impact on their incomes than did female-headed households.

Concerning the effects of Fadama II on the three asset terciles, only those Fadama II beneficiaries in the second tercile increased their incomes significantly more (at $p = 0.05$) than the nonbeneficiaries in that tercile. The percentage and absolute value increases for the second tercile were the largest of the three terciles. This finding indicates that the project had less of an immediate impact on poverty reduction among the poorest households than on others. However, the magnitude of the estimated impact on incomes of the poorest asset tercile is large (45 percent, although it is statistically insignificant, reflecting the high variance of this estimate).[5] Comparable results for the unmatched sample show that the incomes of beneficiaries in the poorest tercile increased significantly by 45 percent, while the middle- and upper-tercile income changes increased but the increase was not significant at 10 percent. Still, the incomes of the poorest asset tercile appear to have been affected less than those of the second tercile, possibly because of the initial investments that the poor had to make to participate in the project. Such investments could have crowded out short-term investments for the poorest, most liquidity-constrained households that could have otherwise increased

[5] The lack of statistical significance of impacts in the estimation subsamples was partly caused by reduced sample size, which depresses statistical power, and does not necessarily mean that Fadama II had no impact in those cases. A larger survey sample would have been required to identify impacts with statistical confidence in such subgroups.

their incomes in the first year of participation. It is likely that beneficiaries in the poorest tercile will see their incomes increase significantly after starting to benefit from their investments in productive assets, which, as discussed above, increased significantly. The significant impact of Fadama II on the unmatched sample could be due to nonproject effects. These results underscore the important role that initial conditions play in benefiting from CDD that targets the poor. Hence in addition to the strong institutions required to address elite capture, initial conditions of wealth also could limit the impact of well-targeted CDD projects on women and the poor. Such beneficiaries may not benefit as much as men and the well-off beneficiaries—at least in the short run.

Here we relay a brief example that illustrates the large impact of the program on beneficiary incomes.[6] A group of 20 women from Adamawa joined Fadama II, raised money to pay for the matching funds, and acquired a milling machine. The average daily income from the milling machine is at least 1,000 nairas, which is 50 nairas per group member or 11,250 nairas per year—assuming the machine works for only five days a week and for 45 weeks per year. This amount is equivalent to a 35 percent increase in income for the female-headed households whose per household income was 35,000 nairas per year before joining Fadama II.

In summary, beneficiaries of Fadama II have realized significant increases in their incomes. Using the PSM and double-difference methods, our results allow us, with considerable confidence, to attribute the income increases among beneficiaries to participation in the project.

However, the impact of Fadama II is different across agroecological zones and asset terciles. The impact of Fadama II on income was not statistically significant in the humid forest and moist savannah zones and across gender, although increases in mean incomes of Fadama II beneficiaries were observed in all cases.[7] Beneficiaries in the lowest and highest asset terciles also did not realize statistically significant different income growth because of participation in the project (although the estimated mean impact was large and positive for the poorest asset tercile). The impacts of the project are not fully captured by this study, because the project had been in operation for only one full year when the survey was done; thus our results do not capture the delayed effects of productive assets, rural infrastructure, and other project interventions. However, the study has collected a good baseline that could be used to conduct follow-up studies to capture the longer-term impacts of the project.

---

[6] Other success stories of Fadama II can be accessed at <http://fadama.org/>.

[7] In the unmatched sample, income of beneficiaries fell, but the change was not significant at $p = 0.10$.

Spillover Effects of Fadama II

We examined the spillover effects of Fadama II by comparing income changes of beneficiaries with those of nonbeneficiaries living in and outside communities participating in Fadama II (Figures 5.3 and 5.4). The results show no significant difference between the income changes of Fadama II beneficiaries and nonbeneficiaries living in the same community. We also compared the changes in income of nonbeneficiaries in Fadama II LGAs with those of nonbeneficiaries outside Fadama II LGAs. Treating nonbeneficiaries in Fadama II as a treatment group (stemming from spillover effects) and nonbeneficiaries outside Fadama II LGAs as the control group, we computed the PSM and matched the two groups. As expected, results show that there was positive spillover of the program on the incomes of nonbeneficiaries in Fadama II LGAs (Table 5.9). Overall, the income of nonbeneficiaries increased by 18 percent because of spillover effects, but the increase was not significant at $p = 0.10$. The largest (and significant) spillover was observed in the moist savannah zone—where the absolute value of ATT of income was greatest.

Figure 5.4 Impact of participation in Fadama II on household income of beneficiaries compared with control groups

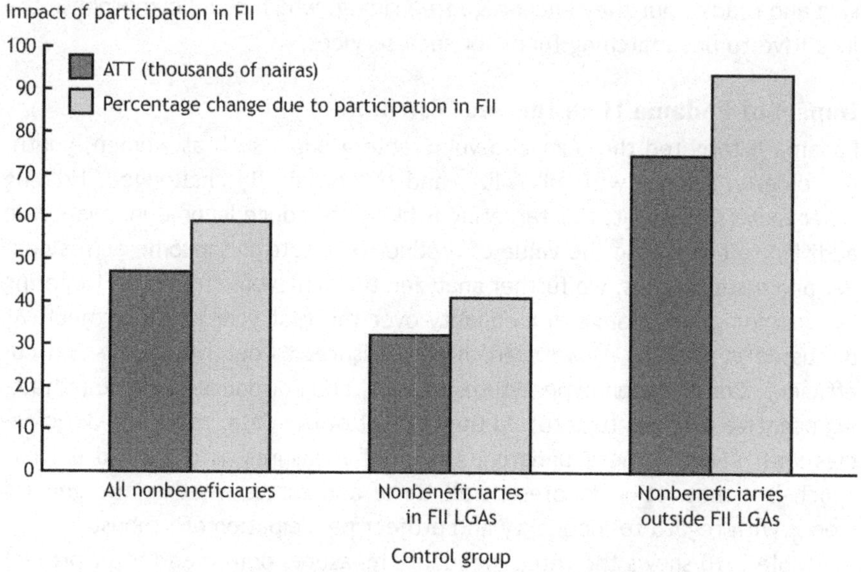

Notes: ATT and the corresponding percentage change refer to the change in income resulting from participation in Fadama II compared with the corresponding group of nonbeneficiaries. Thus these values should not be interpreted as the change in income of the corresponding control group of nonbeneficiaries. ATT, average effect of the treatment on the treated; FII, Fadama II; LGA, local government authority.

Table 5.9 Change in income from spillover effects of Fadama II among nonbeneficiaries in Fadama II LGAs (matched sample)

Characteristic	ATT[a]	Change in income (%)
Nonbeneficiaries in versus those outside Fadama II LGAs	15,309	18
Agroecological zone		
Humid forest	18,931	72
Moist savannah	42,137*	31
Dry savannah	13,748	11
Gender of household head		
Female	7,700	30
Male	17,073	16
Asset tercile		
Tercile 1 (poorest)	15,969	17
Tercile 2	37,186	36
Tercile 3 (wealthiest)	-6,117	-12

Notes: * indicates significance at the 10 percent level. ATT, average effect of the treatment on the treated; LGA, local government authority.
[a]ATT is computed as $[E(Y_1 | p = 1) - E(Y_0 | p = 0)] - [E(Y_0 | p = 1) - E(Y_0 | p = 0)]$.

Such spillover effects enhance the provision of public services (such as extension and roads), but they encourage free riding, which could compromise the incentive to pay matching funds for such services.

Impact of Fadama II on Income Distribution

Fadama II targeted the poor and vulnerable groups, such as women, youth, the elderly, people with HIV/AIDS, and the physically challenged. Holding other factors constant, this targeting is likely to reduce income inequality. In addition to comparing the value of productive assets and income across gender and asset terciles, we further analyzed the achievements of this targeting by examining the change in inequality over the first year of the project. As discussed in Chapter 4, we used three measures to determine the Gini coefficient: Consumption expenditure, income, and normalized income (adjusting negative incomes to zero). In the context of our data, there are deficiencies with each of these underlying welfare measures as discussed earlier, which is a motivation to present all three and assess whether the general trends with regard to inequality and project participation are robust.

Table 5.10 shows the three inequality measures before and after project start for project beneficiaries and for the three categories of nonbeneficiaries also used in earlier analysis. The results show that the Gini coefficient of Fadama II beneficiaries decreased by 9, 24, and 4 percent when using consumption expenditure, income, and normalized income as the welfare indicator,

Table 5.10 Impact of Fadama II on consumption distribution

Treatment type	Measure	Gini coefficient			Negative income (%)	
		Before FII[a]	After FII[a]	Change (%)	Before FII[a]	After FII[a]
	Consumption	0.408	0.370	-9.3		
FII beneficiaries (*n* = 1,281)	Income	1.718	1.302	-23.5		
	Normalized income	0.571	0.554	-3.5	32.4	28.0
Nonbeneficiaries						
	Consumption	0.319	0.325	1.9		
All (*n* = 2,469)	Income	1.746	1.545	-11.5		
	Normalized income	0.546	0.548	0.4	31.7	31.9
	Consumption	0.323	0.317	-1.9	30.2	27.8
In FII LGAs (*n* = 1,240)	Income	1.443	1.226	-14.3		
	Normalized income	0.546	0.541	-1.8		
	Consumption	0.316	0.331	4.8		
Outside FII LGAs (*n* = 1,229)	Income	2.301	2.115	-8.7		
	Normalized income	0.542	0.555	3.7	33.3	36.4

Note: FII, Fadama II; LGA, local government authority.
[a]"Before FII" indicates the year before Fadama II started (October 2004-September 2005). "After FII" indicates the year after the project started (October 2005-September 2006).

respectively. Moreover, the reduction in the Gini coefficient for any one welfare measure is greater in the case of beneficiaries than for any nonbeneficiaries. The last two columns in the table show the percentage of the household observations that have negative incomes. Overall it is quite high, which is why it is inadvisable to rely solely on the Gini coefficient based on normalized income without considering the coefficients based on the other two measures. However, note that the share of negative incomes is similar across participation categories, suggesting that, although this Gini is downward biased, the bias can be expected to be similar across the participation categories, and it is the inequality comparison across these that is relevant.

Despite the imperfection of the income and consumption measures for calculating the Gini coefficient, the robustness of the results across measures and across definitions of nonbeneficiary strongly suggests that the project contributed to the reduction of inequality. This finding is consistent with the results of our productive asset analysis, which showed that the value of group-owned productive assets increased more significantly both in terms of percentage increase and ATT among the poorest asset tercile than among the middle and upper terciles. However, the results are not consistent with the income results in which we observed no significant impact of Fadama II on income for the poorest asset tercile for the matched sample.

We also investigated the impact of Fadama II on income distribution using the coefficient of variation. The income results are more reliable than the Gini coefficient results, which were computed using consumption expenditure data. Such consumption data are less reliable than the income data, because they were collected by asking respondents to recall their consumption expenditures over a span of two years. Table 5.11 reports the changes in the coefficient of variation before and after the start of Fadama II. The

Table 5.11 Coefficient of variation of household income before and after Fadama II

	Coefficient of variation		
Treatment type	**Before FII[a]**	**After FII[a]**	**Change (%)**
All households (*n* = 3,750)	3.28	2.78	-15.2
FII beneficiaries (*n* = 1,281)	2.00	0.90	-55
Nonbeneficiaries in FII LGAs (*n* = 1,240)	0.65	0.66	1.5
Nonbeneficiaries outside FII LGAs (*n* = 1,229)	0.66	0.77	16.7

Note: FII, Fadama II; LGA, local government authority.
[a]"Before FII" indicates the year before Fadama II started (October 2004-September 2005). "After FII" indicates the year after the project started (October 2005-September 2006).

results show that the global coefficient of variation for income decreased by 15 percent. The largest decrease in the dispersion of incomes was observed among Fadama II participants, whose income coefficient of variation fell by 55 percent. The dispersion of incomes for nonbeneficiaries in and outside Fadama II LGAs increased, indicating that inequality increased among nonbeneficiaries.

These results are consistent with those using the Gini coefficients, where we also found increasing inequality among all nonbeneficiaries and those outside Fadama II LGAs. The reductions in income inequality cannot be attributed solely to Fadama II, given that its impact on the incomes of the lowest asset tercile and among female-headed households was not significant. Thus other factors could have contributed to the reductions.

Impact of Fadama II on Advisory Services

Overview

Approaches to and the performance of agricultural extension in Nigeria have been changing over the past few decades (Oladele, Koyoma, and Sakagama 2004). These changes have been driven by many factors, including political and policy modifications, donors, and recently by the participation of NGOs in funding and providing agricultural extension services (Oladele, Koyoma, and Sakagama 2004; Ozor et al. 2007). At present advisory services in Nigeria are largely provided by the Agricultural Development Program (ADP), which evolved from a project funded by the World Bank. The World Bank project started in 1974 and had a broad objective of increasing food production to attain food self-sufficiency (IEG 2001; Oladele, Koyoma, and Sakagama 2004). When the project ended in 1995 with significant success in increasing agricultural production, the federal government adopted its approach and incorporated it into the new ADP. Operations of the ADP are mainly funded by the federal and state governments, and provision of advisory services remains in the hands of public extension workers (Oladele, Koyoma, and Sakagama 2004). The ADP has continued to use the traditional supply-driven approach and has also been characterized by poor funding and less-than-optimally effective advisory services (Ozor et al. 2007). These weaknesses have likely limited the impacts of the ADP on agricultural productivity in the country and on rural development in general.

NGOs and projects have also been providing advisory services and other agriculture-related services (credit services and agricultural input supply). The approaches of the advisory services provided by NGOs and projects have differed, reflecting various focuses and locations in the country. Although the government has allowed and supported NGOs and projects, it has not yet taken bold steps to promote pluralistic advisory services. However, the presence of NGOs and projects has created opportunities for introducing demand-driven advisory services funded by nonpublic sources. Fadama II is one of the projects that provides such services. The project has also introduced the user-fee approach that could help in promoting pluralistic extension services in developing countries (Umali-Deininger 1997). Fadama II beneficiaries contribute 10 percent of the cost of the advisory services they receive. The establishment of a user fee in Fadama II, demonstrating a demand-driven approach, is likely to serve as a good case study for the government to use to design policies for implementing pluralistic extension services in Nigeria and other developing countries. Below we discuss the performance of Fadama II in providing demand-driven extension services and how that has affected provision of production, processing, financial management, and marketing advisory services.

Impact of Fadama II on the Types of Technologies Adopted and Demanded

Table 5.12 shows that a technology used by a large percentage of households surveyed (about one-quarter) was improved crop varieties. Another one-quarter of Fadama II beneficiaries also used financial management technologies, probably because such use was one of the conditions for joining the project. The technology that was demanded by the largest share of households differed across types of household. Fadama II beneficiaries asked for postharvest technologies more than did nonbeneficiaries. The difference in demand for postharvest technologies was significant at $p = 0.10$. This difference could be a reflection of beneficiary demand to make use of the productive assets they had acquired through the PAA component. Surprisingly, nonbeneficiaries demanded financial management technologies more than did beneficiaries. This finding could be the result of having project facilitators who supply beneficiaries with financial management technologies, preempting the need to ask for such technologies.

There was no significant difference between adoption and demand for improved crop varieties. Nonbeneficiary demand for soil fertility management was significantly higher (at $p = 0.05$) than that of beneficiaries (10 percent versus 4 percent). That reflects the limited emphasis of Fadama II on soil fertility technologies. However, the project has addressed soil fertility problems by launching an agricultural input-support component in 2006 (NFDO 2006).

Fadama II beneficiaries also used significantly more livestock management, postharvest handling, financial management, and agricultural marketing services than did nonbeneficiaries. The results suggest that Fadama II

Table 5.12 Adoption and demand for production, postharvest, financial management, and marketing technologies

	Proportion reporting using the technology			Proportion reporting asking for the technology		
Technology	FII beneficiaries (*n* = 621)	All nonbeneficiaries (*n* = 1,107)	Paired test *p* value	FII beneficiaries	All nonbeneficiaries	Paired test *p* value
Improved crop varieties	0.243 (0.017)	0.248 (0.013)	0.840	0.056 (0.018)	0.093 (0.017)	0.170
Soil fertility management	0.126 (0.015)	0.109 (0.001)	0.327	0.038 (0.022)	0.105 (0.027)	0.041**
Livestock management	0.196 (0.016)	0.150 (0.011)	0.013***	0.047 (0.019)	0.041 (0.015)	0.795
Postharvest handling	0.082 (0.011)	0.061 (0.007)	0.103*	0.271 (0.054)	0.160 (0.041)	0.098*
Financial management	0.246 (0.017)	0.061 (0.007)	0.000***	0.013 (0.009)	0.069 (0.030)	0.022**
Agricultural marketing	0.098 (0.012)	0.072 (0.008)	0.059*	0.016 (0.016)	0.059 (0.026)	0.199

Notes: Figures in parentheses are standard errors. Statistics are computed using matched sample only. *, **, and *** indicate significance at the 10, 5, and 1 percent levels, respectively. FII, Fadama II.

support may have given the beneficiaries incentives to use new technologies and may have contributed to the higher incomes that beneficiaries realized.

Table 5.13 shows that the ADP is the major provider of production technologies (improved crop varieties, soil fertility management, and livestock production) for both beneficiaries and nonbeneficiaries. The ADP's focus on providing mainly agricultural production technologies is similar to the pattern of public extension services observed in other developing countries (Qamar 2005). However, it is interesting to note that the State Fadama Development Offices (SFDOs) are the source of production technologies for about 30 percent of the beneficiary households who adopted those technologies.

For both Fadama II beneficiaries and nonbeneficiaries, the sources of postharvest, marketing, and financial management advisory services are mainly NGOs and projects (Table 5.14). As expected, SFDOs are the major source of information for postharvest, financial management, and marketing advisory services for Fadama II beneficiaries. This predominance demonstrates Fadama II's support for such technologies, which Fadama I did not provide. Surprisingly, the SFDOs also provided postharvest, financial management, and marketing technologies to nonbeneficiaries. However, the share of nonbeneficiaries who received advisory services for those technologies from SFDOs was lower than the share of nonbeneficiaries who received the corresponding technologies from the ADP. The results suggest that there is a spillover effect of Fadama II to nonbeneficiaries through these advisory services. Thus free riding is occurring: Fadama II beneficiaries contribute 10 percent of the costs of advisory services, whereas nonbeneficiaries presumably do not contribute.

The results have implications on the user-fee arrangement that Fadama II employs. Collection of user fees from non-Fadama II households could be difficult, because they may not be in organized groups and may not have any form of contract that could facilitate the collection of fees. Additionally, some advisory services are provided using mass media, which makes it difficult to collect fees from those who benefit from such services. Even though Ozor et al. (2007) observed that most farmers expressed willingness to pay for advisory services, payment of user fees by poor farmers who produce low-value crops is a major problem in low-income countries (Qamar 2005), and 100 percent public funding of advisory services for such farmers may still remain the only viable option.

It is also interesting to note that farmer groups and individual farmers are important providers of some advisory services. For example, 25 percent of nonbeneficiaries received agricultural marketing advisory services from fellow farmers. Fadama II has also used radio and television programs to promote various technologies. Radio is an especially important tool for dis-

Table 5.13 Sources of advisory services by type of production technology

	Proportion of households[a]		
Technology / source of information	FII beneficiaries	All nonbeneficiaries	Paired test *p* value
Improved crop varieties	*n* = 149	*n* = 277	
Agricultural Development Program	0.49 (0.03)	0.55 (0.04)	0.271
Individual	0.05 (0.02)	0.07 (0.02)	0.448
Fadama User Association / Fadama Community Association	0.10 (0.03)	0.08 (0.02)	0.463
Farmer association	0.03 (0.02)	0.01 (0.01)	0.043**
Radio/television	0.11 (0.03)	0.04 (0.01)	0.011***
State Fadama Development Office / facilitator	0.31 (0.04)	0.04 (0.01)	0.000***
Ministry of Agriculture and Natural Resources[b]	0.01 (0.01)	0.04 (0.01)	0.065*
Other	0.08 (0.02)	0.08 (0.02)	0.892
Soil fertility management	*n* = 79	*n* = 133	
Agricultural Development Program	0.59 (0.06)	0.78 (0.04)	0.002***
Individual	0.08 (0.03)	0.07 (0.02)	0.897
Fadama User Association / Fadama Community Association	0.13 (0.04)	0.07 (0.02)	0.128
Farmer association	0.05 (0.03)	0.03 (0.02)	0.509
Radio/television	0.03 (0.02)	0.04 (0.02)	0.574
State Fadama Development Office / facilitator	0.36 (0.05)	0.01 (0.01)	0.000***
Ministry of Agriculture and Natural Resources[b]	0.01 (0.01)	0.04 (0.02)	0.261
Other	0.10 (0.03)	0.09 (0.03)	0.902
Livestock management practices	*n* = 128	*n* = 172	
Agricultural Development Program	0.40 (0.04)	0.52 (0.04)	0.040**
Individual	0.06 (0.02)	0.19 (0.03)	0.002***
Fadama User Association / Fadama Community Association	0.12 (0.03)	0.07 (0.02)	0.143
Farmer association	0.05 (0.02)	0.04 (0.01)	0.574
Radio/television	0.02 (0.01)	0.03 (0.01)	0.470
State Fadama Development Office / facilitator	0.32 (0.04)	0.09 (0.02)	0.000***
Ministry of Agriculture and Natural Resources[b]	0.04 (0.02)	0.17 (0.03)	0.000***
Other	0.08 (0.02)	0.13 (0.03)	0.074*

Notes: Numbers in parentheses are standard deviations. *, **, and *** indicate significance at the 10, 5, and 1 percent levels, respectively. FII, Fadama II.

[a]Refers to only those households that adopted the technology and is based on matched sample only.

[b]State-level ministry.

Table 5.14 Sources of postharvest handling, business and/or financial management, and marketing advisory services

Technology / source of information	Proportion of households[a] FII beneficiaries	All nonbeneficiaries	Paired test *p* value
Postharvest handling	*n* = 50	*n* = 66	
Agricultural Development Program	0.41 (0.07)	0.68 (0.05)	0.001***
Individual	0.04 (0.03)	0.17 (0.05)	0.023**
Fadama User Association / Fadama Community Association	0.04 (0.03)	0.07 (0.03)	0.446
Farmer association	0.02 (0.02)	0.00 (0.00)	0.249
Radio/television	0.02 (0.02)	0.03 (0.02)	0.738
State Fadama Development Office / facilitator	0.70 (0.05)	0.13 (0.04)	0.000***
Other	0.04 (0.03)	0.10 (0.04)	0.1964
Business/financial management	*n* = 155	*n* = 66	
Agricultural Development Program	0.12 (0.02)	0.28 (0.04)	0.0003***
Individual	0.01 (0.01)	0.09 (0.03)	0.0013***
Fadama User Association / Fadama Community Association	0.07 (0.02)	0.04 (0.01)	0.3084
Farmer association	0.01 (0.01)	0.02 (0.01)	0.3309
Radio/television	0.01 (0.01)	0.02 (0.01)	0.3309
State Fadama Development Office / facilitator	0.68 (0.04)	0.07 (0.02)	0.0000***
Ministry of Agriculture and Natural Resources[b]	0.00 (0.00)	0.01 (0.01)	0.2145
Other	0.03 (0.01)	0.08 (0.03)	0.0267**
Agricultural marketing	*n* = 62	*n* = 77	
Agricultural Development Program	0.19 (0.05)	0.3 (0.05)	0.1360
Individual	0.06 (0.03)	0.25 (0.05)	0.0032***
Fadama User Association / Fadama Community Association	0.11 (0.04)	0.09 (0.03)	0.6175
Farmer association	0.03 (0.02)	0.00 (0.00)	0.1043*
Radio/television	0.00 (0.00)	0.09 (0.03)	0.0177**
State Fadama Development Office / facilitator	0.62 (0.06)	0.11 (0.04)	0.0000***
Other	0.03 (0.02)	0.12 (0.04)	0.0578*

Notes: Numbers in parentheses are standard deviations. *, **, and *** indicate significance at the 10, 5, and 1 percent levels, respectively. FII, Fadama II.
[a]Refers to only those households that adopted the technology and based on matched sample only.
[b]State-level ministry.

seminating advice on agricultural technologies (Nwaerondu and Thompson 1987). Radio communication is becoming increasingly important in rural areas, where private ownership of FM radios continues to spread. The major challenge in the use of mass media is to ensure that programs are accessible to all listeners by using local languages in rural areas. Some SFDOs, such as those in

Lagos, Ogun, and states in the dry savannah zone, are using local languages in the projects financed by Fadama II.

Factors Affecting Adoption of Technologies

To determine the impact of Fadama II on the adoption of production and postproduction technologies, we analyzed the factors that determine adoption of these technologies, including Fadama II participation as a covariate. Table 5.15 shows that participation in Fadama II significantly increased the propensity to adopt soil fertility technologies, despite the low demand for such technologies by Fadama II beneficiaries reported earlier. This result suggests that the adoption of soil fertility management practices came largely from supply-driven approaches, an observation that is consistent with Qamar (2005). Participation in Fadama II also increased the adoption of business planning, livestock management, postharvest technologies, marketing practices (including marketing intelligence, group marketing, and bargaining), and financial management. In contrast, female household heads were less likely to adopt the last four technologies and practices. These findings on the impact of Fadama II participation were expected, given that Fadama II offered advisory services on financial and marketing issues to beneficiaries as part of its capacity-strengthening activities, and the difference on impact by gender is in line with the common phenomenon that women often have limited opportunity to make use of advanced technologies.

Large families were less likely to adopt postharvest technologies, marketing, and business and financial management technologies, and thus are less likely to be involved in marketing and nonproduction technologies. Consistent with other studies (Scherr and Hazell 1994; Nkonya et al. 2004), education reduces the propensity to adopt labor-intensive technologies (soil and water conservation structures and soil fertility practices). However, education increases the probability of adopting livestock management practices, which are likely to have higher returns to labor, given that demand for livestock products has been increasing in Nigeria and elsewhere in the world because of increasing incomes (Delgado et al. 1999; Ogunyika and Marsh 2006). The value of rainfed land area reduces the probability of adopting improved crop varieties, livestock production, postharvest technologies, financial and business management practices, and marketing technologies.[8] This trend could be due to the adoption of crop varieties and marketing and financial management technologies for large-scale farmers prior to 2004–05, which pre-empted the need to adopt new technologies during the study period (2004–05 through 2005–06).

[8] Rainfed land area is the area of land that is not irrigated and in which crop production depends entirely on rainfall.

Table 5.15 Determinants of adoption of technologies, 2005-06

Variable	All technologies	Crop varieties	Soil fertility	SWC	Livestock breeds	Livestock management	Post-harvest	Marketing	Business plans	Financial records
Fadama II	-0.654	0.553	1.105**	1.459	0.368	0.756*	1.458***	1.951***	1.971*	2.511***
Household size	0.009	0.007	-0.006	-0.009	-0.007	-0.009	-0.023***	-0.019***	-0.021**	-0.033***
Demographic characteristics of household head										
Female	0.173	-0.074	-0.104	-0.183	-0.086	-0.193*	-0.209*	-0.312***	-0.049	-0.375***
Age (years)	0.006*	0.003	0.004	0.004	0.010*	0.005	0.008*	0.000	0.006	0.003
Education (years)	0.012*	-0.008	-0.022***	-0.026**	0.025**	-0.003	-0.005	0.002	0.008	0.005
Value of rainfed land in 2006 (thousand nairas)	-0.002***	-0.001**	13.455	18.646	11.693	-49.544**	-39.028*	-29.554**	-29.280**	-0.0002***
Value of productive assets in 2006 (thousand nairas)	7.965**	3.525	58.138	56.516*	61.743	37.559	15.331***	54.786	126.223	-0.0004*
Value of livestock in 2006 (thousand nairas)	19.812	29.836	19.545	737.247	276.062	139.62	0.419	363.982	281.454	0.0001
Agroecological zone (compared with humid forest)										
Moist savannah	0.711***	0.844***	-0.313**	0.011	0.08	-0.015	0.280**	0.391***	0.432**	0.298***
Dry savannah	0.338***	-0.067	0.052	0.028	-0.119	0.444***	0.085	-0.221*	0.388**	0.177**
Distance to town in 2006	0.001	0.002	0.0003	0.001	-0.004	-0.002	0.001	-0.002	-0.002	-0.009***
Distance to road in 2006	0.0002	0.0004	-0.001	-0.001	-0.008**	-0.001	-0.001*	-0.004***	-0.002***	-0.002***
Constant	-0.598**	-1.197***	-1.275***	-2.060***	-2.163***	-1.422***	-1.879***	-1.428***	-2.227***	-1.579***

Notes: Values are from two-stage weighted regressions. Results from the first-stage estimation (participation in Fadama II subproject) are not reported, because Appendix A reports comparable results. *, **, and *** indicate significance at the 10, 5, and 1 percent levels, respectively. SWC, soil and water conservation.

The value of productive assets owned increased the probability of adopting all technologies, soil and water conservation practices, and postharvest technologies. This finding is consistent with past studies (see the review by Feder and Zilberman 1985) that have observed higher adoption rates among wealthier farmers. However, the value of productive assets is negatively associated with the adoption of financial management skills, which could be due to an adoption of such skills before the start of Fadama II. Surprisingly, the value of livestock does not have a significant impact on the adoption of any of the technologies considered. It was expected that the value of livestock would increase technology adoption—just as for other assets. It is possible that the farmers owning many valuable livestock—like the pastoralists in northern Nigeria—do not employ crop production technologies or improved livestock production technologies, such as livestock breeds.

Adoption of most technologies was generally higher in both the moist and dry savannah zones during the study period. As expected, proximity to roads increased adoption of improved livestock breeds and postharvest, marketing, business, and financial management technologies. Similarly, proximity to a town encouraged adoption of financial management technologies. These results confirm the role that rural services play in helping farmers access production, farm management, and marketing technologies and services. Rural services also encourage the adoption of technologies for the production of perishable products, such as milk and other animal products.

Fadama II has focused on providing postharvest handling, agricultural marketing, and financial management advisory services. However, it is only for postharvest advisory services that beneficiaries have significantly greater demand compared with nonbeneficiaries (at $p = 0.10$). However, participation in Fadama II increased the propensity to adopt soil fertility management, suggesting that the program may have supplied such technology, given that the beneficiaries did not demand it. Nonbeneficiaries reported significantly greater demand for soil fertility management technologies and financial advisory services than did beneficiaries (at $p = 0.05$).

It is not clear why nonbeneficiaries expressed higher demand for soil fertility management technologies than beneficiaries. Perhaps the public extension services provided by the ADP—the major source of advisory services for nonbeneficiaries—provide production technologies, which include soil fertility management practices. The results suggest the need to increase the provision of postproduction technologies—including processing, storage, marketing, and financial advisory services—among nonbeneficiaries. Those services are important for implementing the National Economic Empowerment and Development Strategy framework, which aims to reduce poverty by transforming subsistence agriculture to commercial agriculture (NNPC 2004). The results underline the greater demand for postproduction advisory services (especially regarding postharvest handling) among beneficiaries and the weak demand for production technologies. Thus there is a need to increase the capacity of beneficiaries to demand production technologies to avoid potential land degradation.

Fadama II has had limited impact on the provision of production advisory services. However, the ADP has focused on providing production advisory services using mainly a supply-driven approach. Thus the two projects appear to be complementing each other but use different approaches. The country has used several extension approaches promulgated by donors and projects (Oladele, Koyoma, and Sakagama 2004). As it strives to reform its extension services to be more pluralistic, the government needs to harmonize existing approaches and seek to use those that are complementary rather than conflicting (Oladele, Koyoma, and Sakagama 2004). Complementary approaches will certainly increase the effectiveness of the advisory services. For example, Fadama II has already resulted in experience gained in providing demand-driven nonproduction technologies (postharvest, marketing, and processing technologies), whereas the ADP has inculcated extensive experience in providing production technologies.

However, technologies that require expensive investments with long-term payoffs (such as soil and water conservation structures) may have low demand (Qamar 2005). Therefore, their adoption may initially require the ADP supply-driven approach. It is also important for Fadama II to invest in providing advisory services on production technologies, because the ADP has limited funding to provide such services. Provision of production advisory services will increase the returns from the large investment that Fadama II beneficiaries make when they acquire productive assets. For example, providing advisory services on fish farming could help to increase productivity of that new enterprise.

CHAPTER 6

Conclusions and Policy Implications

In its first year of operation, Fadama II realized significant positive impacts on productive asset acquisition and household income. Using PSM and double-difference methods to control for project placement and self-selection biases, we found that Fadama II dramatically increased the value of group-owned productive assets of the poorest beneficiaries, both in terms of absolute value and percentage. This increase is largely due to the subsidy provided to help finance the acquisition of such assets. The impact of Fadama II on the value of privately owned productive assets was not significant for female-headed households or for beneficiaries in the lowest asset tercile (the poorest beneficiaries). Thus targeting of the poorest individuals through group-owned productive assets did not trickle down to private productive assets, which the program did not directly support. The value of private productive assets increased significantly for the middle and upper asset terciles. This finding is to be expected, given that acquisition of private assets requires investment using the household's own resources, which middle and high income beneficiaries are more likely to afford than are those in the lowest tercile.

Household incomes improved substantially more for Fadama II beneficiaries than for nonbeneficiaries, with an average increase in real income resulting from participation in Fadama II of about 60 percent, well above the target of at least a 20 percent increase that the program set out to achieve in six years for half of its participants. About 42 percent of beneficiaries increased their incomes by at least 20 percent within one year of Fadama II implementation, indicating that the project nearly succeeded in achieving its income goal in its first year of operation.[1]

Comparison of the income impacts of the project across asset terciles shows that the project did not have a statistically significant impact on income among the poorest tercile (although the estimated coefficient was positive) and among female-headed households, despite the large and signifi-

[1] However, as noted earlier, the increase does not control for the influence of other factors that could have contributed to an increase in beneficiary incomes.

cant impacts on group-owned productive assets reportedly available to the women and the poor. These results underscore the important role that initial conditions play in making CDD projects preferentially beneficial to the poor. Hence in addition to elite capture, initial conditions of wealth may also limit the impact of well-targeted CDD projects on female-headed households and the poor. Female-headed households and poor beneficiaries may not benefit as much as male-headed households and the well-off beneficiaries—at least in the short run. A follow-on study is required to assess the medium- and long-term impacts of Fadama II. Such a study will be able to capture the delayed effect of productive asset acquisition, infrastructure, and other benefits accrued from participation in Fadama II. This study was conducted at an early stage of the project and does not capture the delayed impacts of these investments.

The results also suggest the need to help the poor access affordable credit services. The supervision mission and the external medium-term evaluation recommended further reduction of the beneficiary contribution to 10 percent for women and the vulnerable (World Bank 2007a). Even though this recommendation addresses the short-term objective, it is not likely that the approach will be sustainable after the project ends. Affordable rural credit services are the long-term solution for the failure of the poor to be able to pay for productive assets. Fadama II did not involve credit service providers to help beneficiaries pay for their contributions. However, Fadama III, which started in late 2008, has addressed this problem and designed the Fadama Users' Equity Fund, which will provide revolving funds managed by the farmer groups. Loans from this fund will be available for new and existing members. But there is still a need to strengthen the role of private microfinance institutions to provide rural finance services to beneficiaries. For example, credit guarantee and intermediary programs could help poor beneficiaries to borrow and pay for the 30 percent matching funds and other related costs, so that they can benefit both in the short and long terms.

The project had more limited impacts on income in the humid forest and moist savannah zones than in the dry savannah zone. That could be a result of investments in irrigation that beneficiaries in the dry savannah zone demanded over other types of productive assets to address the erratic rainfall in the area. Irrigation investments have a larger impact on agricultural productivity in moisture-stressed areas than in more humid areas.

Fadama II had significant spillover effects in LGAs where the program operated. Compared to households in LGAs that did not participate in the program, the incomes of nonbeneficiaries in Fadama II LGAs increased by 18 percent. However, the increase was not statistically significant. In the moist savannah zone, however, income for nonbeneficiaries in Fadama II increased significantly due to spillover effects. Such spillover suggests greater returns to public investment. These effects are also inevitable for public investments that are hard to limit to program participants. Spillover implies free riding, and it compromises the incentives for beneficiaries to pay matching funds. Mechanisms for introducing user fees for rural roads and for other CDD public investments in which beneficiaries paid matching funds should be explored.

The impact of Fadama II on productive asset acquisition is large and statistically significant across all agroecological zones, asset terciles, and genders. However, the change in the value of productive assets caused by participation in Fadama II was larger and more significant for group-owned productive assets. The dramatic increase in the value of productive assets resulting from participation in the project was mainly caused by the cash transfer from the 70 percent matching funds that the project provides to FUGs. The large cash transfers used to implement this project raises the important question of whether this success story can be replicated in other states.

Three major issues that need to be addressed in scaling up this success story are better targeting of poor and vulnerable groups, finding sustainable methods for promoting development of rural financial services, and increasing the capacity of *fadama* resource users to manage productive assets efficiently. These three issues are interrelated and therefore need to be considered simultaneously.

Over the first year that the project operated, the coefficient of variation of beneficiary incomes decreased by 55 percent, whereas that of nonbeneficiaries increased, suggesting that the project contributed to the reduction of income inequalities. But calculation of this coefficient did not control for other factors that could also affect improvement in income distribution. Additionally, the impact of Fadama II on incomes of households in the poorest tercile and on female-headed households was not significant. Hence there is weak evidence that Fadama II improved income distribution.

Addressing the low capacity of the poor and vulnerable to manage productive assets efficiently calls for increased training and development of complementary services, such as advisory services. One of the components of Fadama II was provision of demand-driven advisory services. The project increased the demand for postharvest handling technologies but did not have a significant impact on the demand for financial management and marketing information (although it did increase adoption of these technologies). Fadama II reduced the demand for soil fertility management technologies, but it increased their adoption, suggesting that the program supplied them or created conditions that led farmers to adopt them. The low demand for soil fertility technologies could also be due to Fadama II's emphasis on providing postproduction advisory services. In its third phase, the program will support agricultural advisory services and input support, and it has mainstreamed sustainable land management. These and other interventions will address the potential land degradation that could result from increased agricultural productivity.

Overall, Fadama II has achieved its goal of increasing the incomes of its beneficiaries in the first year of its operation. The project has also succeeded in targeting the poor and vulnerable in its group-owned productive-asset component, even though that did not appear to significantly increase short-term household incomes and private productive assets among the poorest asset tercile and female-headed households.

The unique feature of Fadama II that may have contributed to the significant impact of the project in a short time is its broad-based approach, which addresses the major constraints limiting the success of CDD projects that address only one or two constraints. This finding has implications on planning poverty-reduction efforts in low-income countries. Given that the poor face numerous constraints, a CDD project that simultaneously addresses many of them will likely build synergies that lead to larger impacts than is possible for a project that addresses only one or two of these constraints. Thus governments and donors need to pool resources and initiate multipronged CDD projects rather than rely on a scattershot approach of instituting numerous isolated projects.

APPENDIX **A**

Probit Regressions of Fadama II Participation (Matched Observations)

Explanatory variable	FII beneficiaries compared with: All nonbeneficiaries: Coefficient	Standard error	Nonbeneficiaries in FII LGAs: Coefficient	Standard error	Nonbeneficiaries outside FII LGAs: Coefficient	Standard error
Gender of household head (1 = female, 0 = male)	0.531***	0.09	0.592***	0.11	0.549***	0.12
Household size	0.021***	0.01	0.021**	0.01	0.023**	0.01
Age of household head (years)	-0.001	0.00	0.010**	0.01	-0.013***	0.01
Education of household head (years of formal education)	0.001	0.01	0.007	0.01	-0.005	0.01
Area of rainfed land (ha)[a]	0.0001*	0.00	0.0001	0.00	0.0001***	0.00
Agroecological zone (compared with humid forest)						
Moist savannah	-0.067	0.12	-0.205	0.14	-0.088	0.16
Dry savannah	-0.039	0.11	-0.06	0.12	-0.121	0.13
Distance to all-weather road[a] (km)	0.0004	0.00	0.0002	0.000	0.005***	0.00
Value of productive assets[a] (million nairas)	-0.432	0.348	-0.642	0.420	-0.284	0.349
Value of livestock assets[a] (million nairas)	0.026	0.045	0.470	0.383	0.043	0.058
Constant	-0.660***	0.23	-0.773***	0.27	0.332	0.30
Sample size (*n*)	966		697		614	
R^2	0.037		0.048		0.097	
Probability > χ^2	0.000***		0.000***		0.000***	
Log likelihood	-606.16		-459.91		-379.97	

Notes: *, **, and *** indicate significance at the 10, 5, and 1 percent levels, respectively. FII, Fadama II; LGA, local government authority.

[a]Quantities reported for the period before the project started.

APPENDIX B

Determinants of Participation in Economic Interest Groups

Variable	Maximum likelihood coefficient
Gender of household head (1 = female)	-0.019
Household size	0.007***
Age of household head (years)	-0.002***
Level of education of household head (years)	0.003**
Area of rainfed land in 2005 (ha)	0.076***
Value of productive assets in 2005 (nairas)	-0.014
Distance to all weather road 2005 (thousand km)	0.005***
Value of livestock in 2005 (nairas)	0.034
Agroecological zone (compared with humid forest)	
Moist savannah	0.006
Dry savannah	-0.005

Notes: Includes all observations. ** and *** indicate significance at the 5 and 1 percent levels, respectively.

APPENDIX C

Comparative Statistics of the Matched and Unmatched Samples

Characteristic	Treated group			Control group		
	Matched	Unmatched	*p*-value	Matched	Unmatched	*p*-value
Human capital						
Household size	10.7	12.5	0.001***	9.0	10.3	0.000***
Gender of household head (1 = female)	0.37	0.39	0.463	0.23	0.28	0.002***
Education of household head (years)	9.7	8.8	0.008***	9.4	8.9	0.06*
Age of household head (years)	42.8	45.2	0.000***	42.4	43.1	0.143
Current use of credit (1 = yes)	0.17	0.15	0.211	0.11	0.08	0.022**
Membership in groups (1 = yes)						
Cooperative group in 2005	0.36	0.23	0.000***	0.27	0.19	0.000***
Cooperative group in 2006	0.35	0.24	0.000***	0.31	0.18	0.000***
Fadama User Group in 2005	0.76	0.85	0.000***	0.56	0.63	0.000***
Fadama User Group in 2006	0.98	0.97	0.854	0.59	0.62	0.088*
Access						
Distance to town in 2005 (km)	9.3	9.4	0.943	7.4	10.1	0.000***
Distance to town in 2006 (km)	9.0	10.6	0.124	9.5	11.2	0.017**
Distance to market in 2005 (km)	4.3	5.8	0.131	3.5	6.1	0.000***
Distance to market in 2006 (km)	3.8	5.6	0.020**	3.2	6.1	0.000***
Use of technologies						
Crop varieties	0.24	0.29	0.083*	0.25	0.22	0.064*
Soil fertility	0.12	0.07	0.004***	0.11	0.08	0.006***
Soil and water conservation	0.04	0.02	0.090*	0.02	0.01	0.336
Livestock breeds	0.05	0.05	0.815	0.06	0.03	0.000***
Livestock management	0.17	0.25	0.002***	0.14	0.13	0.446
Postharvest	0.08	0.06	0.106	0.06	0.05	0.233
Business plan	0.16	0.14	0.309	0.05	0.03	0.024**
Financial records	0.25	0.19	0.032**	0.06	0.03	0.000***
Marketing	0.10	0.10	0.852	0.07	0.05	0.015**
Land endowment (ha)						
Land irrigated in 2005	0.67	0.76	0.401	0.39	0.48	0.198
Land irrigated in 2006	1.2	2.1	0.000***	1.2	1.7	0.137
Land rainfed in 2005	3.5	2.8	0.227	2.4	2.3	0.570
Land rainfed in 2006	4.5	4.6	0.839	3.2	3.9	0.065*

(*continued*)

Characteristic	Treated group			Control group		
	Matched	Unmatched	*p*-value	Matched	Unmatched	*p*-value
Value of productive assets (nairas)						
Group assets in 2005	52,009	24,236	0.022**	5,957	8,749	0.267
Group assets in 2006	358,619	155,841	0.000***	4,692	4,682	0.996
Private assets in 2005	45,072	24,554	0.000***	53,657	35,779	0.000***
Private assets in 2006	26,693	16,331	0.019**	11,142	5,850	0.001***
Household income (nairas)						
Income in 2005	72,256	74,896	0.853	67,181	85,942	0.063*
Income in 2006	93,823	101,469	0.616	83,124	108,215	0.022**

Note: *, **, and *** indicate significance at the 10, 5, and 1 percent levels, respectively.

References

Abadie, A., and G. Imbens. 2002. *Simple and bias-corrected matching estimators.* Berkeley, Calif., U.S.A.: Department of Economics, University of California Berkeley.

——. 2006. *On the failure of the bootstrap for matching estimators.* NBER Working Paper T0325. Cambridge, Mass., U.S.A.: National Bureau of Economic Research. <http://papers.ssrn.com/sol3/papers.cfm?abstract_id=912426>. Accessed March 2008.

Alston, J. M., and P. G. Pardey. 2001. *Attribution and other problems in assessing the returns to agricultural R&D.* Agricultural Economics 25 (2-3): 141-152.

Arcand, J., and L. Bassole. 2007. *Does community driven development work? Evidence from Senegal.* Centre d'Etudes et de Recherches sur le Dévelopment International Working Paper. Clermont-Ferrand, France: University of Auvergne.

Arokoyo, T. 2007. Fadama II beneficiary assessment: Capacity building and conflict resolution component. Report submitted to the National Fadama Development Office, Abuja, Nigeria.

Bardhan, P., and D. Mookherjee. 1999. *Relative capture of local and central governments: An essay in the political economy of decentralization.* Center for International and Development Economics Research Paper C99-109. Berkeley, Calif., U.S.A.: University of California Berkeley.

Bascom, W. R. 1952. The *Esusu:* A credit institution of the Yoruba. *Journal of the Royal Anthropological Institute of Great Britain and Ireland* 82 (1): 63-69.

Berrebi, Z., and J. Silber. 1985. The Gini coefficient and negative income: A comment. *Oxford Economic Papers* 37 (3): 525-526.

Binswanger, H., and S. Aiyar. 2003. *Scaling up community-driven development: Theoretical underpinnings and program design implications.* World Bank Policy Research Working Paper 3039. Washington, D.C.: World Bank. <http://papers.ssrn.com/sol3/papers.cfm?abstract_id=636401>. Accessed January 12, 2010.

Bouman, F. J. 1995. Rotating and accumulating savings and credit associations: A development perspective. *World Development* 23 (3): 371-384.

Chen, C., T. Tsaur, and T. Rhai. 1982. The Gini coefficient and negative income. *Oxford Economic Papers* 34 (3): 473-478.

Cleaver, F. 1999. Paradoxes of participation: Questioning participatory approaches to development. *Journal of International Development* 11 (4): 597-612.

Conning, J., and M. Kevane. 2002. Community-based targeting mechanisms for social safety nets: A critical review. *World Development* 30 (3): 375-395.

Dasgupta, A., and V. Beard. 2007. Community-driven development, collective action and elite capture in Indonesia. *Development and Change* 38 (2): 229-249.

Dehejia, R., and S. Wahba. 2002. Propensity score matching methods for non-experimental causal studies. *Review of Economics and Statistics* 84 (1): 151-161.

Delgado, C., M. Rosegrant, H. Steinfeld, S. Ehui, and C. Cloubois. 1999. *Livestock to 2020: The next food revolution*. Food, Agriculture, and the Environment Discussion Paper 28. Washington, D.C., Rome, and Nairobi, Kenya: International Food Policy Research Institute, Food and Agriculture Organization of the United Nations, and International Livestock Research Institute.

Dongier, P., J. van Domelen, E. Ostrom, A. Rizvi, W. Wakeman, A. Bebbington, S. Alkire, T. Esmail, and M. Polski. 2001. Community-driven development. In *A sourcebook for poverty reduction strategies*. Vol. 1, ed. J. Klugman. Washington, D.C.: World Bank.

Duflo, E., R. Glennerster, and M. Kremer. 2006. *Using randomization in development economics research: A toolkit*. Working Paper 138. Cambridge, Mass., U.S.A.: Harvard University, Center for International Development.

Duflo, E., S. Mullainathan, and M. Bertrand. 2004. How much should we trust difference-in-difference estimates? *Quarterly Journal of Economics* 119 (1): 249–275.

Farrington, J., and R. Slater. 2006. Introduction: Cash transfers: Panacea for poverty reduction or money down the drain? *Development Policy Review* 24 (5): 499–511.

Faye, I., and J. Sutherland. 2006. Fadama II supervision mission: May 14–30, 2006. Monitoring & evaluation and readiness for impact evaluation report. Abuja, Nigeria: World Bank. Mimeo.

Feder, G., and D. Zilberman. 1985. Adoption of agricultural innovations in developing countries: A survey. *Economic Development and Cultural Change* 33 (2): 255–298.

FEWSNET (Famine Early Warning System Network). 2006. *Nigeria food security outlook*. <www.fews.net/nigeria>. Accessed March 2008.

FOS (Federal Office of Statistics). 2004. Nigeria living standard survey 2003/04. Abuja, Nigeria. Photocopy of draft report.

Galasso, E., and M. Ravallion. 2005. Local knowledge vs. local accountability? Decentralized targeting of an anti-poverty program. *Journal of Public Economics* 89: 705–727.

Gbenga, O. 2007. Fadama II beneficiary assessment terminal report. National Fadama Development Office, Abuja, Nigeria. Photocopy.

Gillespie, S. 2004. *Scaling up community-driven development: A synthesis of experience*. Food and Nutrition Division Discussion Paper 181. Washington, D.C.: International Food Policy Research Institute.

Greene, W. 2003. *Econometric analysis*. 5th ed. Upper Saddle River, N.J.: Prentice Hall.

Heckman, J., H. Ichimura, and P. Todd. 1998. Matching as an econometric evaluation estimator. *Review of Economic Studies* 65 (2): 261–294.

Heckman, J., H. Ichimura, J. Smith, and P. Todd. 1998. Characterizing selection bias using experimental data. *Econometrica* 66: 1017–1099.

IEG (Independent Evaluation Group). 2001. *Agricultural development projects in Nigeria*. Washington, D.C.: World Bank. <http://lnweb18.worldbank.org/oed/oeddoclib.nsf/DocUNIDViewForJavaSearch/FE7BA13642E3E0D7852567F5005D85CF#top>. Accessed January 12, 2010.

Imbens, G. M., and J. M. Wooldridge. 2008. *Recent developments in the econometrics of program evaluation*. NBER Working Paper 14251. Cambridge, Mass., U.S.A.: National Bureau of Economic Research.

Khwaja, A. 2001. *Can good projects succeed in bad communities? Collective action in the Himalayas.* Cambridge, Mass., U.S.A.: Harvard University Press.

Kleimeer, E. 2000. The impact of participation on sustainability: An analysis of the Malawi rural piped water scheme program. *World Development* 28 (5): 929–944.

Kohl, B. 2003. Democratizing decentralization in Bolivia: The law of popular participation. *Journal of Planning, Education and Research* 23 (2): 153–164.

Labonne, J., and R. Chase. 2008. *Do CDD projects enhance social capital? Evidence from the Philippines.* World Bank Policy Research Working Paper 4678. Washington, D.C.: World Bank.

Manor, J. 1999. *The political economy of democratic decentralization.* Washington, D.C.: World Bank.

Mansuri, G., and V. Rao. 2004. Community-based and -driven development: A critical review. *World Bank Research Observer* 19 (1): 1–39.

Maziya-Dixon, B., I. Akinyele, E. Oguntona, S. Nokoe, R. Sanusi, and E. Harris. 2004. *Nigeria food consumption and nutrition survey 2001-2003.* Report submitted to the Ministry of Agriculture and Rural Development, Nigeria. Ibadan, Nigeria: International Institute of Tropical Agriculture.

Mosse, D. 1997. Colonial and contemporary ideologies of community management: The case of tank irrigation development in south India. *Modern Asian Studies* 33 (2): 303–338.

NBS (National Bureau of Statistics). 2007. Nigeria living standard survey 2003/2004. Abuja, Nigeria.

NFDO (National Fadama Development Office). 2006. *Poverty reduction through increased productivity and empowerment.* Abuja, Nigeria: NFDO Project Coordination Unit.

Nkonya, E., J. Pender, P. Jagger, D. Sserunkuuma, C. Kaizzi, and H. Ssali. 2004. *Strategies for sustainable land management practices and poverty reduction in Uganda.* Research Report 133. Washington, D.C.: International Food Policy Research Institute.

NNPC (Nigerian National Planning Commission). 2004. *National Economic Empowerment and Development (NEEDS).* Abuja, Nigeria.

Nwaerondu, N., and G. Thompson. 1987. The use of educational radio in developing countries: Lessons from the past. *Journal of Distance Education* 2 (2): 43–57.

Ogunyika, E., and T. Marsh. 2006. Nigerian demand for meat. *Journal of International Food & Agribusiness Marketing* 18 (1):125–149.

Okonjo, K. 1979. Rural women's credit system: A Nigerian example. *Studies in Family Planning* 10 (11–12): 326–331.

Oladele, O., O. Koyoma, and J. Sakagama. 2004. Africa in search of extension systems: Experience from Nigeria. *Food, Agriculture and Environment* 2 (1): 276–280.

Ostrom, E. 1990. *Governing the commons: The evolution of institutions for collective action.* New York: Cambridge University Press.

Ozor, N., A. Agwu, N. Chukwuone, M. Madukwe, and C. Garforth. 2007. Cost-sharing of agricultural technology transfer in Nigeria: Perceptions of farmers and extension professionals. *Journal of Agricultural Education and Extension* 13 (1): 23–37.

Platteau, J. P. 2004. Monitoring elite capture in community-driven development. *Development and Change* 35 (2): 223-246.

Qamar, K. 2005. *Modernizing national agricultural extension systems: Practical guide for policy makers of developing countries.* Rome: Food and Agriculture Organization of the United Nations.

Ravallion, M. 2005. *Evaluating anti-poverty programs.* Policy Research Working Paper 3625. Washington, D.C.: World Bank.

Rosenbaum, P. R., and D. B. Rubin. 1983. The central role of the propensity score in observational studies for causal effects. *Biometrika* 70 (1): 41-55.

Salter, A. J., and B. R. Martin. 2001. The economic benefits of publicly funded basic research: A critical review. *Research Policy* 30 (2): 509-532.

Scherr, S., and P. Hazell. 1994. Sustainable agricultural development strategies in fragile lands. Environment and Production Technology Division Discussion Paper 1. Washington, D.C.: International Food Policy Research Institute.

Schoen, P., U. Hassan, and P. Okoli. 2002. *Resource use conflict management study in Nigeria.* Abidjan, Ivory Coast: African Development Bank.

Smith, J., and P. Todd. 2001. Does matching overcome LaLonde's critique of non-experimental estimators? *Journal of Econometrics* 125 (1-2): 305-353.

Stata. 2007. Stata statistics/data analysis, version 10. College Park, Tex., U.S.A.

Stich, A. 1996. Inequality and negative income. *Statistical Methods and Applications* 5 (3): 297-305.

Umali-Deininger, D. 1997. Public and private agricultural extension: Partners or rivals? *World Bank Research Observer* 12 (2): 203-224.

World Bank. 2003. Project appraisal document for the second national Fadama development project. Abuja, Nigeria. Photocopy.

——. 2007a. Medium-term review: Independent assessment on the second national Fadama development project (September 2004-January 2007). Abuja, Nigeria. Photocopy.

——. 2007b. Joint Federal Government of Nigeria/World Bank supervision mission, 6-16 February 2007. Contribution to Mission Aide-Mémoire related to Monitoring and Evaluation (M&E) and Management Information System (MIS). Washington, D.C. Photocopy.

——. 2007c. *World development indicators, 2007.* Washington, D.C.

Yahaya, M. 2007. Beneficiary impact assessment of the second National Fadama Development project: Communication and advisory services component. Report submitted to the National Fadama Development Office, Abuja, Nigeria. Photocopy.

About the Authors

Ephraim Nkonya is a senior research fellow in the Environment and Production Technology Division of the International Food Policy Research Institute, Washington, D.C.

Dayo Phillip is a professor of agricultural economics at the Nasarawa State University, Keffi, Lafia Campus, Nigeria.

Tewodaj Mogues is a research fellow in the Development Strategy and Governance Division of the International Food Policy Research Institute, Washington, D.C.

John Pender is a senior economist in the Economic Research Service of the U.S. Department of Agriculture, Washington, D.C. He was previously a senior research fellow in the Environment and Production Technology Division of the International Food Policy Research Institute, Washington, D.C.

Edward Kato is a research analyst in the Environment and Production Technology Division of the International Food Policy Research Institute, Washington, D.C.

Index

Page numbers for entries occurring in figures are suffixed by *f*, those for entries occurring in notes by *n*, and those for entries in tables by *t*.

Made in the USA
Monee, IL
04 November 2026

67379771R00056